BRITISH MOTOR FISHING VESSELS

JOHN McWILLIAMS

AMBERLEY

First published 2018

Amberley Publishing
The Hill, Stroud
Gloucestershire, GL5 4EP

www.amberley-books.com

British Library Cataloguing in Publication Data.
A catalogue record for this book is available from the British Library.

ISBN 978 1 4456 7863 4 (print)
ISBN 978 1 4456 7864 1 (ebook)

Origination by Amberley Publishing.
Printed in Great Britain.

INTRODUCTION

Helped by improved markets for their fish thanks to the spread of the railways during the nineteenth century, the fishing fleets expanded and often prospered. Beautiful and functional sailing fishing boats appeared in ports all around Britain, including Brixham trawlers, Essex smacks, Lowestoft dandies, Scots fifies and zulus, Manx nickeys, Cornish luggers and many others.

Steam trawling began on the Humber in the 1880s and developed rapidly during the next decade. In 1897 the first English steam drifter was built and soon hundreds of these jaunty craft fished from East Anglia and ports right along the East Coast of Scotland.

In the early years of the twentieth century, the internal combustion engine began to be installed in small local fishing boats. Motors reduced the fishermen's back-breaking labour and greatly increased efficiency. In the West Country, this trend was supported by Fishery Officer Stephen Reynolds. This accelerated further during the First World War, with the help of motor loans from the Ministry of Agriculture & Fisheries, and much of the fishing fleet was fitted with motors. Although Lowestoft and Brixham built sailing smacks to replace those sunk by U-boats during the war, for most inshore fishermen the way ahead was clear; for Britain's smaller fishing ports, the future was with the motor fishing boat.

The sailing fishing boats of the British Isles are well documented in several excellent books, but now is the time to tell the story of the motor fishing boats that replaced them, as many kinds of local boats have gone or are disappearing fast. The ring netters and drifters of Scotland, the beach boats of Norfolk, Suffolk and Devon, and the pilchard boats and long liners of Cornwall are already history. Modern boats are built of fibreglass and steel, and are likely to be much the same whether they work from Shetland or Devon.

One problem with a book like this is what to include and what to leave out. The straightforward answer would be just to describe wooden boats, but the fine steel seiners and twin riggers built for Scotland in the past forty years are certainly typical Scots boats, as were the steel steam drifters built in the decade before the First World War. The shallow-draught steel cockler/prawners of Boston and King's Lynn are local boats, designed for fishing in the Wash. The Cygnus 32-foot boats built of fibreglass in the 1970s were just as typically Cornish as the classic local luggers that preceded them; therefore, at the risk of offending purists, I have included them.

British Motor Fishing Vessels

For much of the twentieth century, the herring seasons played a major role for hundreds of vessels. The Scots and East Anglian drifters began their fishing off Shetland in May, and a month or two later they were working from Aberdeen, Fraserburgh and Peterhead. By August the Scots drifters were landing their catches at Whitby and Scarborough. In the autumn the drifter fleets were based in Lowestoft and Great Yarmouth for Home Fishing, which were once the greatest herring fisheries in the world.

The ring net fleet developed rapidly between the two world wars. The Clyde boats often migrated from their home waters to the Hebrides for the winter, running their catches across the Minches to land at Mallaig. The summer saw the ring netters working out of Peel in the Isle of Man, supplying the local kipper houses and landing their fish at Portpatrick or Ardglass in times of glut. Other ringers migrated to work from Seahouses, Bridlington, Whitby or Scarborough. The local herring seasons enlivened ports all around the British Isles.

Migrating Lowestoft and Yarmouth drifters came west each spring to fish the mackerel shoals from Newlyn, where they were joined by local boats and rugged little wooden drifters, the *malamoks*, from Douarnenez in Brittany, which often outfished them.

For much of the twentieth century the Cornish drifters came west for the summer pilchard season to Mount's Bay and worked around the Wolf Rock – 'going to the Wolf', as it was called. Each autumn saw them gathering in Plymouth Barbican as the shoals moved eastwards.

With the development of pair trawling and purse seining from the 1960s, a period of rapid change began in the pelagic fisheries. The herring was replaced in importance by the mackerel, which is now caught off the north of Scotland rather than Cornwall. The hundreds of drifters and ring netters of the past have been replaced by a very few, huge, high-tech midwater trawlers based in Shetland, Fraserburgh and Peterhead. Their catches are measured in hundreds of tonnes. In Shetland, about nine pelagic vessels have taken the place of 400 drifters.

Many other fisheries were seasonal, and trawlers too went where the fish were. Each spring saw the Lowestoft and Milford trawlers arrive on the Trevose grounds to fish for soles, while Brixham boats came west to work the 'Wolf' grounds. Scalloping has greatly prospered in recent years, with scallopers migrating from the West Country to the Seine Bay or off the Welsh coast.

Other fishing methods have changed. For example, beam trawling and twin rigging have replaced side trawling, and cuttlefish, largely destined for Spain and Italy, has become an important target of the beamer fleet.

Much of the Scots twin rigger fleet now concentrates on prawns rather than white fish. Recent years have seen some of their number working

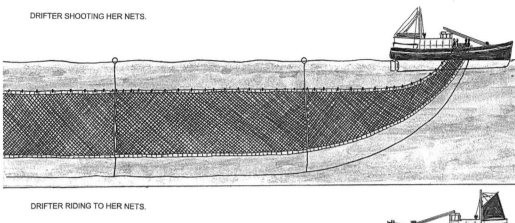

DRIFTER SHOOTING HER NETS.

DRIFTER RIDING TO HER NETS.

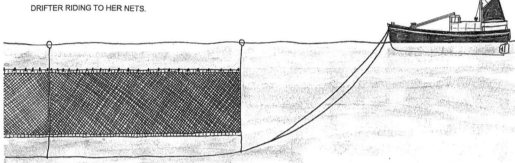

Drifter at work.

RING NETTING

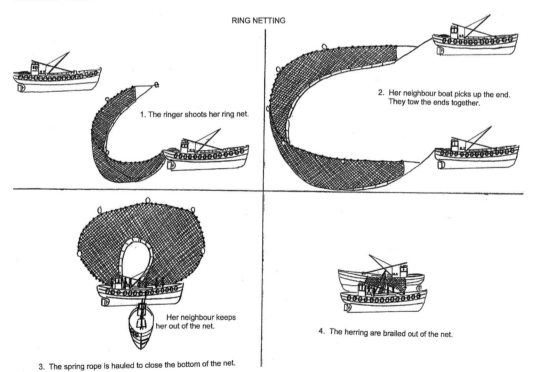

1. The ringer shoots her ring net.

2. Her neighbour boat picks up the end. They tow the ends together.

Her neighbour keeps her out of the net.

3. The spring rope is hauled to close the bottom of the net.

4. The herring are brailed out of the net.

Ring netting.

British Motor Fishing Vessels

Jones's Bank, west of the Isles of Scilly, and landing their catches at Newlyn for transport to Scots processers.

Gill nets have replaced long lines and the market for net-caught hake has been developed. Huge quantities of pots or nets are worked by static gear boats, though a successful shellfish skipper recently commented the following about a crabber working 3,000 pots: 'If you can't earn a living with 300, you're in the wrong job.' There is much more emphasis on high quality, rather than quantity. Catches are handled with care and boxed and iced at sea, and many vessels have chilled fishrooms. West Country ring netters land top-quality sardines from chilled sea water tanks and crabbers are fitted with vivier shellfish tanks to keep their catch alive.

Local boats were usually built by local boatyards. Each boatbuilder had a detailed knowledge of a skipper's requirements, not just for the boats that were suitable for a particular fishery, trawling, drifting, lining, ring netting or seining, but vessels that could work from the local harbour. A tidal harbour, afflicted by heavy ground swell, required a boat capable of surviving these rugged conditions. Recent trawlers designed to fish from Looe, a tidal port, have shallower draught than those intended to work from nearby Plymouth. This means that a boat designed to sail from a deep water dock will not be suitable for working from a beach.

With the greatly increased efficiency of modern boats, their number has declined. Harbours that once thrived with vibrant fishing industries are now empty or devoted to leisure craft. Boats have now standardised, so now is the time to describe some of the many motor fishing boats that once worked around our coasts, before they are forgotten.

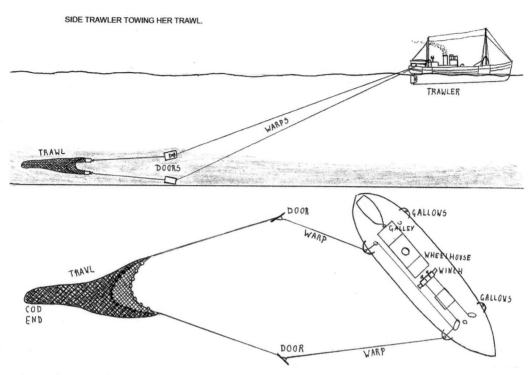

Side trawler at work.

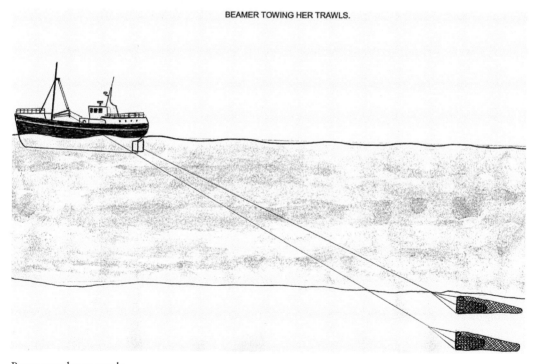

Beam trawler at work.

THE BOATS

MOUNT'S BAY PILCHARD BOAT

Mount's Bay sailing pilchard boats from Porthleven, Newlyn and Mousehole were miniature square-sterned luggers of about 25 to 30 feet in length. There was a cuddy forward with sliding doors on the bulkhead. When fishing, the foremast was lowered down in the scottle (a long rectangular slot in the foredeck). There were narrow side decks and an after locker. A low coaming went around the working part of the boat and in the middle of the boat was the netroom, covered by removable hatches.

With the arrival of motors, mainly during the First World War, a bigger class of motor pilchard boat came into use. These decked boats were about 32 to 40 feet long with the engine room aft, which was usually accessed by a hatch behind the wheelhouse. In the middle of the boat were the netroom and fishroom, with waterways on either side. The cabin was forward with its coal stove, bunks and lockers that were usually accessed by a companion way with a sliding hatch on the port side. The scottle was replaced by a tabernacle for the foremast. There were two engines, one on the centre line and another on the port side; typically a 26 hp Kelvin in the centre and a 7 hp quarter engine. These reliable Kelvins started on petrol and then changed over to run on paraffin, which was much cheaper. The boats were fitted with acetylene gas lights, sometimes called hats due to their distinctive shape. A few new motor pilchard boats were built, among them the *Girl Joyce* PZ 156 and *Penzer* PZ 324 of Mousehole, but most Mount's Bay motor pilchard boats were converted luggers, and were often recognisable by the hole cut in the rudder for the centre propeller, since they were built without propeller apertures. When they were motorised, these boats were usually fitted with wheelhouses and gaff mizzen sails, which set flatter than the traditional lug sail. The mizzen kept the boat steady, head to wind, when she was riding to her fleet of nets.

Surprisingly, many sailing pilchard boats were sawn in half, lengthened, and raised to make them into motor boats. These included the *Bonnie Lass* PZ 306, *We'll Try* PZ 198 and *Faithful* PZ 302 of Mousehole, the *Emulate* PZ 208 and *Girl Lillian* PZ 319 of Newlyn, the *Boy George* PZ 576 of Porthleven and the *Nazarene* SS 114 and *Ocean Gift* SS 60 of St Ives.

The *Bonnie Lass*, for example, was built in St Ives in 1905 with a length of 28 feet. She returned to St Ives in 1924 and was rebuilt with a length of 40 feet and 11 feet beam. She fished until the 1960s, as did many of these veterans.

The *Faithful* was originally an open 25-foot lugger. When rebuilt in 1921 she was a decked motor pilchard boat, 36 feet long with 10.4 feet beam and 7 hp and 15 hp Kelvins. There can't have been much of the original boat left; it might have been easier just to start from scratch and build a new boat.

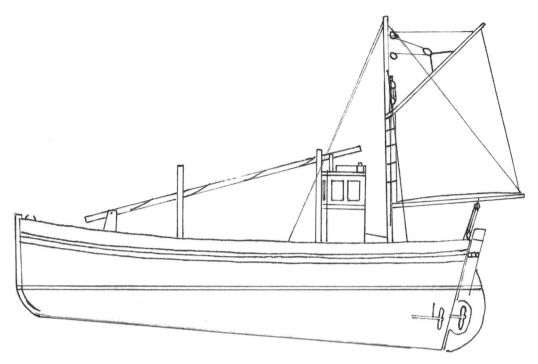

Above: Mount's Bay pilchard boat.

Below: Pilchard boats at Mousehole in 1939. Left to right they are: *Harvester* PZ 126, *Gleaner* PZ 36, *Guide Me* PZ 416, *Bonnie Lass* PZ 306 and *Faithful* PZ 302.

British Motor Fishing Vessels

CORNISH LONG LINER/DRIFTER

Among the earliest large Cornish motor boats built before the First World War were the *Treryn Castle* PZ 190 of Newlyn, built in 1910, and the *Ben My Chree* PZ 545 of Mousehole, built in 1912. Both of these boats worked the local drift net and long-line fisheries and sailed to Ireland for herring. The *Ben My Chree* was well-known in Ireland, the Isle of Man, East Anglia and at Mallaig, in the west of Scotland.

The drawing shows the *Our Lizzie* SS 55, which was built by Olivers of Porthleven for the Penberthy family of St Ives and was registered in 1920. Her details were as follows: length 40.2 feet, beam 13.8 feet, depth 6.2 feet and 24.42 tons. Registered length was from the stem to the rudder post. Since the *Our Lizzie*'s stern extends about 4 feet beyond the rudder post, her overall length is around 44 feet. She was fitted with a 26 hp Kelvin and a 24 hp Gardner in her engine room, which was forward, in front of the netroom and fishroom. Her cabin, aft, was equipped with a coal stove for cooking and a boiler to drive her steam capstan and line hauler, so it must have been an inferno in summer. Since the engines were fuelled by petrol and paraffin, the engines and steam boiler were often at opposite ends of the boat, because of the danger of inflammable vapour. Indeed, several boats were lost to fire. The *Our Lizzie* fished the usual seasons; mackerel drifting in the spring, pilchard driving from Newlyn during the summer and herring driving from St Ives in the autumn. It was quickly realised that the Cornish motor boats could not compete with the Lowestoft steam drifters that came to Newlyn on their westward voyage for mackerel every year, and soon the local boats replaced mackerel drifting by long-lining.

The later Cornish motor fishing boats all had transom sterns. The *Our John* SS 64 was built for the Barber family of St Ives in 1926 by Gilbert & Pascoe of Porthleven. The *Our John* never went mackerel drifting, where footlines were worked on the nets, so she was not fitted with a capstan. Her line hauler was powered by a petrol engine below deck. Her engine room was forward; next came the fishroom, then the netroom, and her cabin was aft. Her length was 44 feet, beam 13.4 feet and depth 5.9 feet. She was originally powered by two 26 hp Kelvins, but one of these was later replaced by a 24 hp Petter diesel.

A successful herring drifter, the ketch-rigged motor boat *Ben my Chree II* was built for Mousehole by Kitto of Porthleven in 1912. She fished the herring all around the British Isles, from Great Yarmouth, Mallaig, Castlebay (Barra), the Isle of Man and Ardglass. In the photograph below, there are empty long-line baskets on her netroom hatch. She has shot her lines and is heading into Mousehole for the night. Forward is her steam line hauler, or jinny. Two of the crew stand by her steam capstan, which was used for hauling the footlines of her herring nets. The pipe right forward is from her steam boiler. Amidships is her engine exhaust and in front of the wheelhouse is the cabin stove pipe. She was sold to Ireland in 1935 and renamed *Howth Head* D 164, and worked a long career out of Howth.

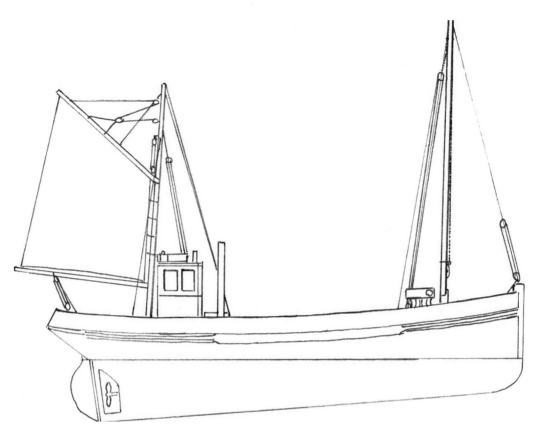

Cornish long-liner/drifter.

Ben My Chree II PZ 545. (Photograph © Penlee House Gallery & Museum, Penzance, Gibson Archive)

British Motor Fishing Vessels

ST IVES GIG

From the late nineteenth to the early twentieth century, an important St Ives season was for herring. This was fished by the smaller class of decked luggers – the pilchard boats, which were from 30 to 37 feet long. These were very seaworthy, but with their high freeboard they often lost herring out of their nets. The shallow-draught rowing and sailing gigs could often go to sea and get in a quick shot while the luggers were still aground. These clinker-built gigs were about 24 to 28 feet long, were pulled by four oars and were rigged with a dipping lug foresail and sprit mizzen. Several of these clinker gigs were lost by overloading or stress of weather, including the *Jabez* in 1878, the *Golden Light* in 1899, the *Fortitude* in 1900, the *Maggie* in 1908 and the *Lily and John* in 1910.

After the First World War these clinker gigs were gradually replaced by almost forty carvel-built motor gigs. The first was the *Glorious Peace* SS 37, which was registered in October 1919. The motor gigs were 30 to 40-foot-long open boats. They were strongly built on sawn frames. The two engines, often a 13 hp Kelvin in the centre and a 7 hp on the quarter, were in a cambered engine house aft, which went from one gunwale to the other. The crew had to climb over this to get to the fore part of the boat. There were up to four thwarts forward and one aft to support the mizzen mast. The boats were shallow-draught, drawing almost nothing forward. Their registration documents describe them as luggers rigged with foresail, mizzen and jib with auxiliary motor engines. Several of them, including the *St Eia* SS 165, *Our Boys* SS 150, *Annie* SS 157 and *Glorious Peace,* were fitted with centre boards, but these were later removed. Most of them were built in the 1920s, although the last, the *Sweet Promise* SS 95, was built in 1947.

Most gigs were progressively modified, with the freeboard raised, wheelhouses and gaff mizzens being fitted, foredecks built, thwarts removed and capstans fitted for crabbing and trawling. Unlike the decked motor boats, which were painted a sombre black, the gigs were usually pale blue or white with black tops. Their usual seasons comprised crabbing in the summer, herring drifting in the autumn and trawling over the winter. Some went round to Newlyn for the summer pilchard driving.

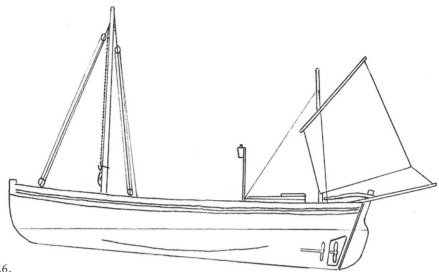

St Ives gig, 1926.

St Ives gig, 1946.

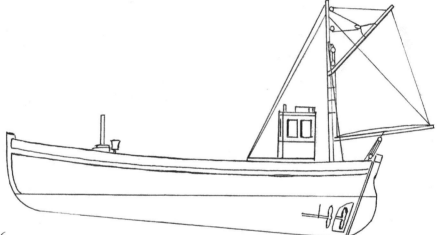

St Ives gigs *Our Girls* SS 131, *Our Boys* SS 150, *Thrive* SS 148 and *Silver Spray* SS 28 in the 1930s.

British Motor Fishing Vessels

CORNISH COVER

Until the arrival of motors during the First World War, many of Cornwall's coves – Port Isaac, Priest's Cove (St Just), Sennen, Porthgwarra, Penberth, Prussia Cove, Cadgwith, Coverack, Mullion, St Mawes, Durgan, Portscatho, Portloe and Gorran – had fleets of lug- or sprit-rigged crabbers, often known as covers (cove-ers). These boats sailed out to their crabpots and then took the foremast down and hauled their gear under oars. The Crab & Lobster Fisheries Report of 1876 shows that several small coves were packed with boats: twenty-four at Gorran, twenty-six at Portloe, thirteen at St Mawes and twenty-two at Sennen. By the early twentieth century these numbers had increased, and about 300 boats crewed by 600 fishermen worked from Cornwall's coves.

Typically these boats worked about forty-five crab pots. At Prussia Cove they worked three strings of fifteen, and at Cadgwith six strings of six or seven. Fishermen made their own dome-shaped withy pots, except at St Ives, where crabpots made of strong wire were introduced to cope with the area's heavy ground swell. Many of the boats were very small, only 16 feet long at Gorran Haven. Sennen Cove is more exposed than Gorran so its boats were bigger, from 18 to 22 feet long. Many of the earlier boats were clinker-built but by the early twentieth century most seem to have been carvel. They had to be shallow-draught to be easily hauled up and launched. This could affect their sailing qualities so some were fitted with centre boards.

Some boats were fitted with petrol engines, but the 3½ hp Kelvin petrol/paraffin engines were popular. Until the arrival of reliable small diesels in the 1960s, many boats retained a pair of oars 'just in case'. The fitting of motors to the cover fleet meant that they could work much more gear. By the 1950s a crabbing crew would make about 150 crab pots over the winter. This resulted in many fewer boats, but there were still about ten crabbers working from Cadgwith. In the 1930s, capstans, known as dollies, were introduced to haul the gear. A drive from the fore end of the engine went under the bottom boards to the bow. Here, half an old lorry back axle was used to turn the shaft through 90 degrees and drive the vertical capstan.

Many former sailing boats had long careers under power. The 17-foot-long *Ladysmith* PZ 33 was built for Porthgwarra in 1900. Sold to Sennen in 1922, she had a 3½ hp Kelvin fitted. In 1928 she was sold to Mousehole, where she worked until the 1960s. The 20-foot *Aphrodite* PZ 188 was built at St Just, in Roseland, in 1926 for Richard and John Chapple of Penberth. In 1939 she was sold to Newlyn and renamed *Coral*. When the Newlyn pilchard boat *Dashing Spray* was tragically lost with her two crew in 1942, the *Coral* went out and recovered one of her nets and 160 stones of pilchards. As SS 38 she had yet another stage to her career as a very successful hand-line mackerel boat from St Ives.

Traditional wooden covers were still being built in the 1970s. The 25-foot-long *Cornish Light* was built at Gweek Quay in 1975 for Arthur Williams and Timothy Goddard of Cadgwith. Powered by a 22 hp Saab diesel, she was fitted with a mechanical bilge pump, a Sea Winch capstan/line hauler and a Petterson net hauler. Cadgwith's present fleet of fibreglass covers are fitted with modern electronic and navigational aids.

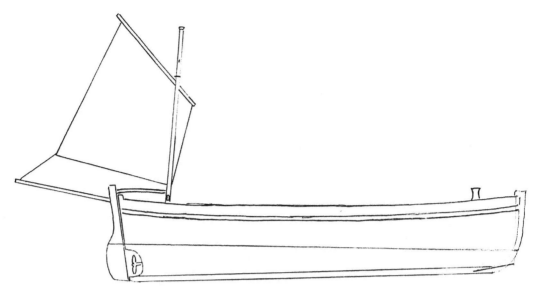

Cornish cover.

Hauling withy pots off Mullion.

British Motor Fishing Vessels

75-FOOT MFV TRAWLER

During the Second World War, the Royal Navy required a fleet of small craft to act as auxiliaries and liberty boats for its warships. Four designs of motor fishing vessels (MFVs) were built, based on already proven Scots fishing boat designs and intended to be sold for conversion to commercial fishing vessels after the war. The MFVs were of 50-foot, 65-foot, 75-foot and 97-foot length overall.

The 75-foot MFVs were very popular in Scotland, where many were adapted to seiner/drifters; some of them well-known record breakers. Many also came to West Country ports, where they were converted for trawling. The 75-foot MFV had a narrow beam of 19 feet 8 inches, and her draught was 5 feet 6 inches forward and 9 feet 6 aft. They were underpowered by 160 hp Lister diesels. They were already fitted with trawl winches, so converting them for trawling was not a major job.

The Stevenson company of Newlyn bought many 75-foot MFVs and had them refitted to its own requirements by local boatbuilders J. Peake & Sons. Their bulwarks were raised and they were equipped with trawl gallows. Later, they were fitted with raised foredecks to keep their decks dry. Their original engines with no reduction gears were gradually replaced by more powerful 240 hp and 320 hp Listers and 320 hp Kelvins. Among the 75-foot MFV trawlers that sailed from Newlyn were the *Trevarth* PZ 189, *Jacqueline* PZ 192, *W&S* PZ 193, *Roseland* PZ 194, *William Stevenson* PZ 195, *Trewarveneth* PZ 196, *Sara A Stevenson* PZ 253 and *Anthony Stevenson* PZ 331. Some of them fished on for half a century with the help of several major refits.

Their wheelhouse equipment was also progressively modified, and by the 1970s included radars, echo sounders, radio telephones and Decca Navigators. The arrival of the Decca was a huge boost for trawlermen, as accurate navigation enabled them to pinpoint wrecks and other fasteners on the seabed that would have destroyed their trawls.

They were successful boats and though they might be seen as modest vessels when compared with today's hefty beamers and netters, they were a big step up from Newlyn's 50-foot liners. They also began the move away from the traditional drift net and long-line fisheries, as described by a local newspaper in 1949:

Deep sea trawling is taking the place of long lining, pilchard, herring and mackerel fishing, which have been the main types of fishing for many years.
Pioneers of deep sea trawling at Newlyn have been Messrs W Stevenson and Sons who have five former MFVs converted into trawlers----their craft have all done well in the new venture, comparing favourably with the Belgian and French trawlers which visit the port.

These vessels worked the local grounds around the Wolf Rock and the Trevose Ground off Padstow, mainly for soles. They also discovered new fishing grounds. At that time, trawling was new to many Cornish fishermen, and the 75-foot MFVs provided a steep learning experience.

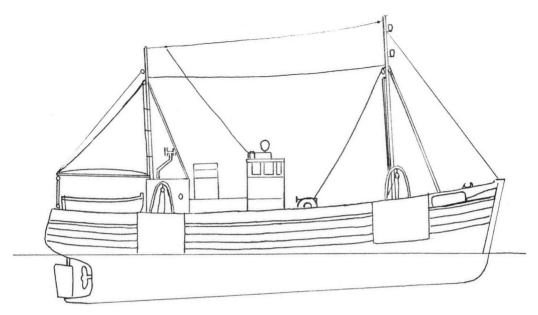

75-foot MFV trawler.

Trevarth PZ 189 at Newlyn in 1960. Inflatable liferafts had yet to appear and she carries a lifeboat aft, behind the galley. Alongside the South Pier is one of Everard's coasters (which were nicknamed 'yellow perils' from their hull colour), loading roadstone from Penlee Quarry.

British Motor Fishing Vessels

CYGNUS GM 32

In 1974 the G. Percy Mitchell boatyard at Portmellon, near Mevagissey, completed the 40-foot-long *Trazbar* SS 104 and *Louise Joanne* SS 106 for St Ives. These fully decked boats were a big advance on previous half-decked craft. Shortly afterwards, their designer, Gary Mitchell, designed the Cygnus GM 32-foot boat to be built in fibreglass by Cygnus Marine of Penryn. These boats were hugely successful; ideal for the winter hand-line mackerel fishery and summer crawfishing with tangle nets. Many were built and they were the forerunners of a range of bigger netters.

Like the *Trazbar* and *Louise Joanne*, they were fully decked. Until their arrival, it was not thought feasible to deck a boat as small as 32 feet long. The GM 32 was 32 feet long overall and 30 feet 2 inches at the waterline, with 11 feet 6 inches beam and 4 feet draught aft. The woodwork to make the moulds for these boats was produced at the Portmellon yard.

The popularity of the design can be seen from the numbers built, which included the *Crimson Arrow* PW 185 for Boscastle, the *Bri Al An* FH 234 for John Cock of Flushing, near Falmouth, the *Tyack Mor* SS 216 for the Plummer family of St Ives, the *Kendore* SS 245 for Ken and David Brian of St Ives, the *Southern Comfort* FY 276, the *Emma Goody* FH 323 for Alun Davies of Flushing, and the *Rockhopper* FH 328. Skipper Robbie Curtis promptly landed a record shot of 1,060 stone of hand-line-caught mackerel in the *Rockhopper*.

Typical equipment for a Cornish Cygnus GM 32 was a 100 hp Ford Sabre diesel with 2:1 reduction gear, a Seawinch capstan, a Graphete echo sounder, a Westminster VHF radio, a Decca Mk 21 Navigator, a Seascan radar and a set of gurdies for winding up her strings of mackerel. The pedestal hauler was for hauling crawfish nets and the aft gantry arrived soon afterwards. This gantry was not usually intended for trawling; rather, it was to raise the foot of the mizzen mast so there was unobstructed space for shooting nets over the stern. Both of these features can be seen in the drawing.

The first of the series to be completed was the *Emma Goody*, after her chocolate brown hull was displayed at the Catch 75 national fishing exhibition. She was fitted with two double hydraulic mackerel gurdies and one manual. With this outfit, it was expected that her three crew could work five lines. When not engaged in mackerel fishing, her skipper, Alun Davies, worked the local oyster beds in his traditional gaff-rigged working boat. (The use of powered vessels is banned in this fishery.)

The cost of a GM 32, ready to fish, was estimated at £16,000, which was about £4,000 cheaper than the equivalent wooden boat. They were immensely strong; the owner of one of the St Ives boats used to carry a piece of fibreglass in his pocket and invite acquaintances to hit it with a hammer!

The Cygnus GM 32 first appeared in 1975, and a number of them are still at work.

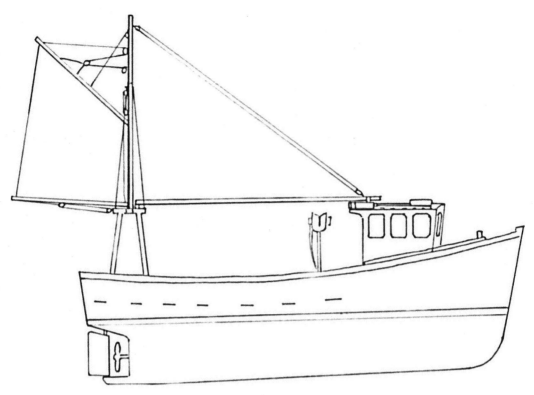

Cygnus GM 32.

The 40-foot St Ives netter *Trazbar* SS 104 was designed by Gary Mitchell and was built by the Percy Mitchell boatyard at Portmellon in 1974. The Cygnus 32 design was derived from the *Trazbar* and *Louise Joanne*. (Photograph Sam Bennetts, courtesy of St Ives Archive)

British Motor Fishing Vessels

CORNISH NETTER

During the 1960s and '70s, netting for crawfish, monkfish and ray took off, and, along with the winter hand-line mackerel season, brought prosperity and optimism to many Cornish ports. Many new wooden and fibreglass boats were built by local yards. However, the crawfish nets devastated the stock, and by the 1980s some fishermen had given up netting. Others diversified into nets for dogfish, hake, turbot, pollack, ling and cod. Craw nets had a 30 cm mesh, while hake nets, which were worked all year round, had a mesh of 4⅞ inches. This has recently been increased to 5½ inches, which is more selective and catches only bigger fish. Wreck nets are worked from January to March to catch pollack, ling and cod, which live in old wrecks. They have a mesh of 16 cm.

In recent years trammels have been reintroduced for turbot, which have a high value. Trammels have been used at least since Tudor times. They consist of three parallel nets all together on the same ropes, with the centre net made of small mesh, and the outer walls of large mesh. When a fish swims into the centre net it carries it through the outer net to make an escape-proof pocket.

These nets are joined together in tiers of 100, except for wreck nets, which are shot in smaller groups. As the netter steams ahead, her nets run out from the net pound aft. They lie on the seabed with anchors, buoy ropes and dans at each end. The nets are hauled by a rubber-covered hydraulic roller – the hauler – which is usually forward on the starboard side. The fish are removed and the nets are hauled back into their pound to be shot again. This used to take two men, but is now done by a machine called the net stacker. These nets cannot be worked in strong tides, so netters work during the weaker neaps. This means that two weeks' wages must be earned in one week's work. Seals scavenge the netters' catch, patrolling the tiers of nets and taking a bite out of each fish.

Many fine netters in the 40-foot range were built in the 1980s. Newquay built a very impressive fleet, including the *Celtic Mor* SS 83 for Trelawney of Cornwall, the *Guiding Light* PW 377, the *Lamorna* SS 28, the *Regina Maris* PW 57, the *Pearn Pride* PW 62, the *Atlanta* PW 182 and the *Trevose* PW 64.

But as the fishery expanded, bigger boats were needed to fish further offshore. Most of these were second-hand from Scotland, and some came from France. They included the 54-foot *Pilot Star* PZ 188 for Newlyn and *Harvester* FH 198 for Falmouth in 1987, the ex-Grimsby seiner *Poul Nielsen* GY 370 for Padstow in 1988, and the ex-Breton *Ar Bargeergan* PZ 287 and *Ajax* AH 32 in 1990. Some new boats were built, including the varnished 50-foot *L&T Britannia V* FH 121 from Nobles of Girvan for Mevagissey in 1986, the *Ocean Spray* PZ 41 from Toms of Polruan in 1990, the 60-foot fibreglass *Sowena* PZ 14 in 1990, the steel *Berlewen* PW 1 for Padstow and the *Silver Dawn* PZ 1196 for Newlyn in 2002. Although the numbers of large netters in Cornwall have declined, the fishery is currently prospering and Cornish hake has a high reputation. All modern large netters are covered with shelter decks and their net pounds are enclosed. They are safe boats, but need to be, as they work far from land, and often in poor weather.

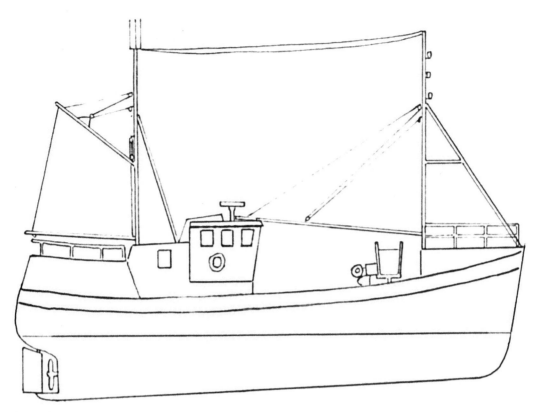

Cornish netter.

Netter *Govenek of Ladram* PZ 51 in 2015. Her working deck is completely covered by a shelter deck, as is her net pound aft. She still sets a traditional mizzen sail to keep her steady on her nets.

MEVAGISSEY TOSHER

The Mevagissey tosher was a one-man boat, designed for plummeting for mackerel in the spring and summer. The bigger luggers were often laid up during the plummeting season, and were fitted out again in the autumn for lining for dogfish, or for herring drifting from Plymouth. During the 1920s and '30s the Mevagissey and Looe luggers came to Newlyn for the summer pilchard driving in Mount's Bay and around the Wolf Rock.

A plummet was a cone-shaped lead with a homemade bright metal spinner. A spinner could only catch one fish at a time, but a tosher would tow three lines – two from poles on each side and one over the stern. There was a short line fastened to each of the lines that were towed from the poles, which enabled them to be hauled in. The tiller was let go when a line caught fish so that the port side propeller would drive the boat round in a circle, hopefully keeping her on the bunch of mackerel.

Many toshers were built in the 1920s by local boatbuilders Roberts, Frazier & Percy Mitchell. In his *A Boatbuilder's Story*, Percy Mitchell recalls Roberts building the *Mavis, Swan, Ena, Chu Chin Chow, Verona, Nancy, Rosana* and *Morvina* while he was an apprentice there. Percy Mitchell's own yard at Portmellon built seven toshers for the 1927 season alone, and Fraziers built an equal number. Among those built by Percy Mitchell were the *Sea Belle, Melody, Pauline, Cresta* and *Charmaine*. In the 1920s these boats cost £53 10s. In 1936, Fraziers were building toshers at £65.

The tosher's length was 20 feet, with a beam of 7 feet and a draught aft of 2 feet 6 inches to 3 feet 6 inches. Their beam was carried well aft and their length was limited to 20 feet as longer boats paid higher harbour dues. They were carvel-built with pine planks on closely spaced Canadian rock elm timbers, which were steamed into place. The keel, garboards and bilge planks were of English elm. The transom, stem and stern posts and floors were of oak.

Toshers were usually powered by 3½ hp Kelvin petrol/paraffin engines, or sometimes 6 hp engines. The fuel tank was in the foredeck and there was an aft locker. The tiller worked through a hole in the transom. There were three thwarts, and the mast was stepped behind the forward thwart. Its sail was stowed with the gaff against the mast, with the gaff jaws facing upwards. They carried about 15 to 18 cwt of iron ballast.

In recent times some toshers have been rebuilt as leisure sailing boats, with lead keels and large gaff rigs. They are reported to be exhilarating to sail.

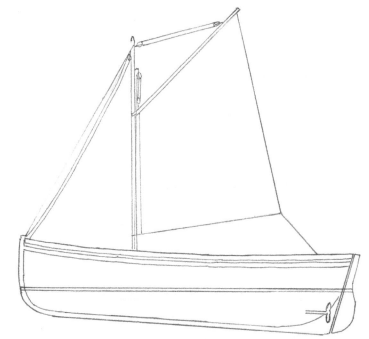

Right: Mevagissey tosher.

Below: Mevagissey toshers
Sea Trout FY 256, *Luneda*
FY 327 and *Nipper* FY 144.
They have distinctive half-
car-tyre fenders over their
legs. Another row of toshers
can be seen against the quay
in the background.

British Motor Fishing Vessels

LOOE LUGGER

In the early years of the twentieth century, the Mount's Bay and St Ives fishermen were in crisis. They concentrated on the deep water mackerel fishery and their sailing luggers were soon outclassed by the visiting Lowestoft steam drifters. However, the pragmatic East Cornish fishermen of Looe, Polperro and Mevagissey concentrated very successfully on the pilchard fishery. As a result, their fleets were fitted with motors before the First World War. A 1913 report into fitting engines in the Cornish boats showed that thirty-one of the thirty-six first-class boats at Looe already had engines fitted.

In the spirit of post-war optimism, several fine motor luggers were built for Looe in the local yards: the *Kathleen* FY 297 and *Eileen* FY 310 in 1920; the *Our Daddy* FY 7, *Dorothy* FY 19 and *I.R.I.S* in 1921; the *Seagull* FY 408 in 1922; the *Progress* in 1924; *Janie* FY 227 in 1925; and *Girl Vine* FY 88 in 1930. The *Emma* FY 299 was built in Porthleven, as were the Mevagissey luggers *Liberty* FY 317, *Boy John* and *Emblem* FY 325. There were already many fairly new boats in the fleet, which had been built as luggers and fitted with engines. Three large counter-sterned boats, the *Forget Me Not* FY 269, *Swift* FY 405 and *Cornishman* FY 122, were built, but they seem to have been too big for Looe and were sold to Newlyn as long-liners.

The Looe motor boats were generally bigger than Mevagissey's miniature luggers and, except for the well-known *Our Daddy*, had their engines forward, exhausting through the side of the boat. Next came the fishroom and netroom, and the cabin was aft. Surprisingly, some of them were built without a propeller arch, like the *I.R.I.S.* She had three engines, and so all three propellers were on the port side.

The *Janie* FY 227 was built by Arthur Collings for R. & W. Pengelly to replace a previous *Janie* FY 371, which was sunk in collision between the Eddystone and Rame Head in December 1924 during the Plymouth herring season. With an overall length of 44.4 feet, her keel was 38 feet long, her beam 12.8 feet and her draught was 5.5 feet. She was fitted with a Kelvin 15 hp engine in the centre and a 13 hp engine on the port side, and carried two lug sails.

The 43-foot *Seagull*, built for Emily Atkinson, had a 33 hp engine and was unusual in having her wheelhouse offset to port. She was carefully maintained by her crew of yachties (fishermen who spent much of the year crewing the big yachts). In May 1938 they shipped a hand capstan and came to Newlyn for the mackerel season.

Since the East Cornish luggers were mainly intended for pilchard and herring drifting, they had very wide hatches, which were suitable for working drift nets, and narrow side decks. When a boat was fitted with a wheelhouse in West Cornwall, she was also fitted with a gaff mizzen, but the Looe men persisted with lug mizzen sails. One problem existed, which was, when the mizzen was taken down, where was it to be put? A frequent answer was to prop one end of the yard on top of the wheelhouse. The outrigger was sometimes propped up there as well.

The Looe and Polperro fishermen used boxes for their long lines, while Mevagissey had small baskets to fit its miniature luggers. The lines were hauled by the jinny, which was driven by a petrol engine in a box on deck. The jinny had two counter-rotating grooved wheels. In the 1950s and '60s the Looe luggers dispensed with the foremast and were fitted with gaff mizzens. Their deck lights were strung from a length of pipe that went from the bow to the wheelhouse.

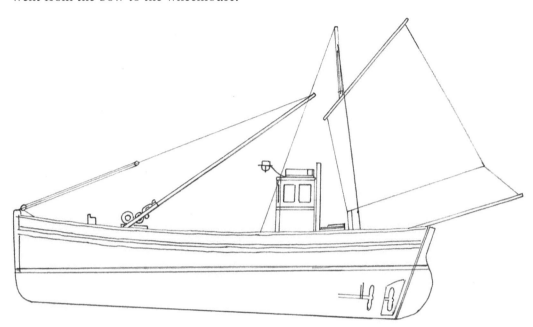

Looe lugger.

The *Janie* was sold to Newlyn in 1935. Though registered PZ 30, she is still a very typical Looe lugger. She has a 'leg of mutton' (triangular) mizzen sail. (Photograph K. Brown collection)

British Motor Fishing Vessels

LOOE QUATTER

At Looe the inshore long-lining grounds were called the Quat. From the late 1930s a fleet of very successful half-decked boats known as quatters, which were about 31 to 37 feet long, were built to work these grounds. Under the long foredeck was the cabin right forward, usually with its coal stove, and then the engine room with two engines – one driving the centre propeller, and the smaller engine intended for the quarter propeller on the port side. The wheelhouse, which was open at the back, went right across the boat, with just narrow side decks on either side. This wheelhouse, or canopy, sheltered the whole boat. There was a lug mizzen aft to keep the boat steady while riding to pilchard nets. The netroom was in the middle of the boat.

The quatter evolved over several years. Between the two world wars, several East Cornish boats were built with the cabin and engine room forward, including the 32-foot-long *Dessie,* built for Luther Pengelly in 1925, and the 34-foot *Manxman,* which was built in 1934. Both these craft had their wheelhouse aft. The little 30-foot *Sea Bird* FY 251 of Polperro, built in 1926, was fitted with a wheelhouse forward, but this was of the traditional telephone box type and did not go right across the boat. With a full-width wheelhouse, the *Endeavour II,* built in 1937, was 31 feet 10 inches long with a beam of 9 feet 6 inches and had a 3-foot 9-inch draught. She was powered by a 30 hp Lister diesel. Sold to St Ives in 1947, she had a long career as a tripper boat before reverting to netting in the late 1960s. Another veteran was the *Silver Searcher,* which was sold to Newlyn skipper Harry Blewett and had a useful career pilchard driving as the *Aronwyn II.* St Ives fishermen thought she was a much more sensible and roomy boat than their gigs.

The well-known *Endeavour* FY 369 was built at East Quay, Looe, by Arthur Collings and was launched by being pushed over the quay on a high spring tide in April 1947. Solidly built of Columbian pine planks on English oak frames, she was 35 feet 3 inches long with a 10-foot 6-inch beam. She had a 30 hp Lister centre engine and a 16 hp engine on the quarter. When fishing declined in the 1950s, and much of the Looe fleet was left to run shark angling trips for holidaymakers, the *Endeavour* could often be seen with pilchard nets aboard. In 1951 the H. Pearn yard completed the quatter FY 268, which had a length of 36 feet, an 11-foot 3-inch beam and a 3-foot 3-inch draught. She was powered by a 21 hp Lister in the centre and a 16 hp on the quarter. Planked with larch and pine on oak frames, she cost £2,100, and was intended for pilchard driving in the winter and long-lining in the summer.

The Looe quatters were much admired and widely copied all around Cornwall and the West Country both for fishing and the tourist industry. For example, the *Sunlit Waters,* a very typical quatter, was built for trawling for Thomas Harvey of Torquay in 1956. More unusual was the cruiser-sterned 36-foot *Polaris,* which was built by Percy Mitchell of Portmellon for Looe in 1950. Fitted with two Dorman diesels, her sound was distinctive. She was one of the first Cornish boats to be fitted with an echo sounder for fish-finding.

The quatters were the mainstay of the Looe fleet, and among them were the *Lady Betty* FY 456, the *Silver Spray* FY 11, the *Ella* FY 440, the *Tethra* FY 116 and the *Genesha* FY 14. When fishing was in decline, many of them earned a living as tripper boats for much of the 1950s and '60s, but when the winter hand-line mackerel seasons took off in the 1970s, they proved ideal for the job, and many of them landed record shots.

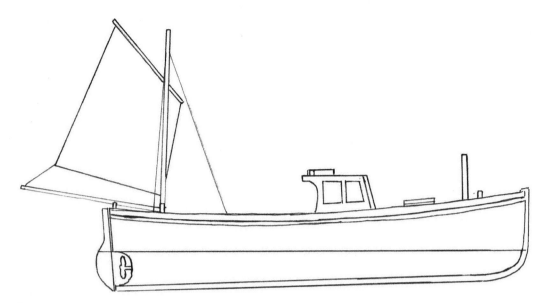

Looe quatter.

Built in 1937, the Looe quatter *Endeavour II* was sold to St Ives after the Second World War and was renamed *Gay Dawn* SS 94.

LOOE TRAWLER

In the 1970s and early '80s, Looe was one of the leading ports for the Cornish winter hand-line mackerel season. With the destruction of the mackerel shoals by the industrial fleet and the rapid decline of the crawfish stocks, which were depleted by netting, several Cornish ports quickly declined. Looe, however, got its second wind, and built a whole fleet of modern inshore stern trawlers. Its new fish market developed a reputation for high-quality white fish from its day boats, regenerating the port.

Some trawlers were built of wood by local yards and others of fibreglass. They were of similar layout, with the cabin forward, then the engine room and the insulated fish room aft. The trawl winch was behind the forward wheelhouse and there was a gantry aft for getting up the trawl. They were fitted with net drums aft.

In January 1980 the *Paravel* FY 369 was completed by the local Gerald Pearn yard for Skipper Mike Soady. She was 40 feet long with a 15-foot beam and a 6.3-foot draught, and was powered by a 290 hp Volvo Penta engine. She was fitted with a Redifon radar, a Koden echo sounder, Neco autopilot, a Decca Navigator and Plotter, and a Sailor radio receiver and VHF radio. (The Looe luggers of a previous, but not too distant generation had a complete navigation outfit of a compass, a log line to record the distance run and the skipper's watch.) Unusually, the *Paravel* was fitted with a stern ramp for hauling her trawl. Her net drum was fitted just aft of her trawl winch.

In January 1985 the 44-foot-long *Maxine's Pride* FY 48, powered by a 325 hp Volvo Penta, was completed for Skipper Gordon Cairns. Like the *Maxine's Pride*, and many of the others, the 35-foot-long *Danvic of Looe* FY 444 was designed by Gary Mitchell and built at the Toms Yard of Polruan. The *Danvic of Looe,* built in 1986, was just under 10 metres long to avoid many of the rules that affected bigger boats. Built for David and Justin Bond, she was powered by a Gardner 172 hp diesel. Built of iroko planking on oak frames, she had a beam of 14 feet and draught of just 5 feet 6 inches. Looe is a tidal harbour, so draught is kept to a minimum.

Others built at this time included the 38-foot *Bilander* FY 589, which was built by Toms in 1981 for Skipper Ernie Curtis, the *Kingfisher of Looe* FY 17, which hailed from the same yard in 1983 for Skipper Lewis Butters, the 40-foot *Galatea* FY 97, which was built of fibreglass by Cygnus Marine for Frank and Mike Pengelly in 1984, the *Palatine* FY 149, which was built by Toms for Mike Soady in 1985, and the *Innisfallen* FY 46, which was built in 1988. Others included the *Cazadora* FY 614, the *Briagha Mara* FY 293 and the *Natalie* FY 602.

Most of these boats were later fitted with shelter decks. The number of trawlers at Looe has declined since the 1980s but, unlike many other fishing ports, it is still active. More recent trawlers have been built of fibreglass or steel.

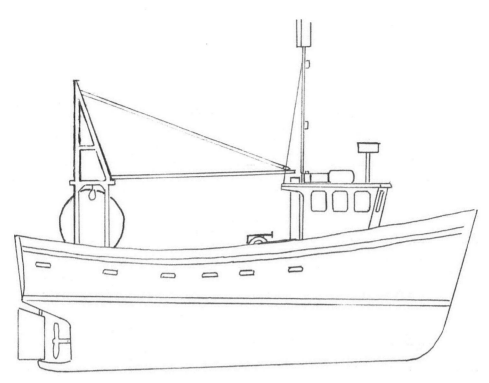

Looe trawler.

Innisfallen FY 46 has a net drum aft. Built in 1986 for Richard & Charlie Butters of Looe, she is still going strong.

British Motor Fishing Vessels

WEST COUNTRY RING NETTER (PURSER)

Although modern high-tech ring netting is a recent and very successful development in Cornwall, ring netting for pilchards was first tried in 1947. At that time, pilchards were caught by the ecological but labour-intensive method of drift netting. In November 1947 the Ministry of Agriculture & Fisheries' research vessel *Onaway* LT 358 and hired Scots ring netter *Hope* began trials with the ring net. Although there were problems, it was proved that a suitable net could be effective.

In 1950 the Mousehole boats *Renovelle* PZ 107 and *Mark H. Leach* INS 207 took up ring netting and were successful in the autumn fishery, where the shoals were more concentrated. Later, their net was modified to a miniature purse seine, which, unlike the Scots ring net, could be worked by one boat. (The Cornish purse seine is, for some reason, always called a 'ring net', and the boats 'ring netters'.)

The West Cornish boats *Sweet Promise* SS 95, *Girl Renee* SS 78, *Couer de Lion* PZ 74, *Renovelle* PZ 177 and *JBS* SS 17 tried purse seining for herring at Dunmore East in Ireland. The *Sweet Promise* and the 50-foot MFV *Renovelle* were very successful, but after a trade deal with South Africa allowed it to undercut the Cornish industry, the pilchard fishery declined.

In the 1990s, Skipper Martin Ellis of Cadgwith bought the *Renovelle*'s net and tried ring netting in his 24-foot cove boat *Samantha Rose*, landing to Nick Howell's Pilchard Works at Newlyn. Finding the *Samantha Rose* too small, he bought the 30-foot *Penrose*, and was so successful in her that she sank with a 450-stone shot near St Michael's Mount. Undaunted, he replaced her with the 40-foot steel *Prevail* PZ 883 from Ireland.

Pilchards were very successfully rebranded as Cornish sardines. Pilchards were just-tolerable tinned fish in tomato sauce, whereas sardines are an upmarket product, rich in health giving oils, grilled in the sunshine with a glass of white wine and part of the European holiday experience.

As ring netting was developed by local skippers, such as Stefan Glinski in the *Pride of Cornwall* SS 87 (and later the *White Heather* SS 33), Sam Lambourne of the *Prue Esther* PZ 550 (and more recently the catamaran *Lyonesse* PZ 81) and others, the markets expanded and major supermarkets became good customers. Andrew Lakeman of Ocean Fish went to Brittany to observe the Breton sardiners and had the *Resolute* FY 119, and more recently the 46-feet *Asthore* PZ 182 and *Mayflower* PZ 187, built by Buccaneer Boats. Ring netting is now an important part of the local economy, with ringers working from Plymouth, Mevagissey and Newlyn from July to February. Although they are modern boats, tradition is strong. FY 119 was the number of the famous Lakeman family lugger *Ibis*. The *Lyonesse*, *Asthore* and *Mayflower* were well-known Newlyn long-liners.

Most ring netters are purpose built of fibreglass. Some revert to netting outside the sardine season. They are very complex and expensive boats, fitted with sonars, modern navigational aids and side thrusters to stop them getting mopped up in their nets. The net is usually hauled with a triplex hauler of three hydraulic rollers. The fish are pumped out of the net with a fish pump and kept in perfect condition in chilled seawater tanks. One ringer can easily land as much as half a dozen drifters.

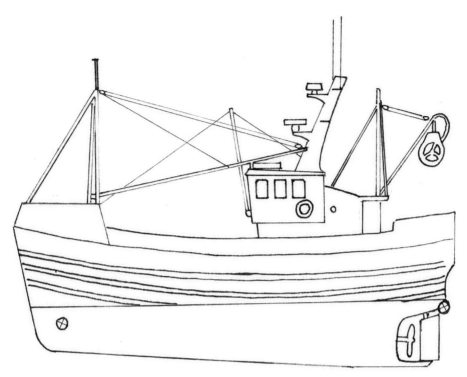

West Country ring netter (purser).

The ring netter *Asthore* PZ 182, landing her sardines with her brailer in 2015. Her net with its purse rings is beside her deckhouse, with her triplex hauler above.

British Motor Fishing Vessels

SCALLOPER

Scallops were fished by the Essex sailing smacks in the nineteenth and early twentieth centuries. In the West Country, scalloping has, like beam trawling and netting, taken off since the 1970s. Scallopers were converted from other kinds of fishing boats; many were ex-French inshore trawlers and sardine boats from Brittany, which were available cheaply. Some of them were near their sell-by date and had short careers. In the late 1970s, during the mackerel boom, some of these boats made headlines when they came in to Falmouth with their dredges full of rotting mackerel that had been slipped by the industrial vessels when they caught more than they could process. These scallopers were modest little craft with engines of up to 120 hp and as few as three dredges a side, which worked the grounds east of the Lizard in Veryan and Gerrans Bay. A 50-foot boat would tow six dredges a side.

Scallops spawn during the summer and mature at three years. They live in a shallow pit in the seabed and filter feed. They are caught by the scallop dredge, which is a rectangular bag towed from a triangle-shaped steel frame. The bottom of the bag is made of steel rings and the top of strong netting. At the bottom of the entrance to the bag is a toothed bar, which scoops the scallops into the dredge as it is towed along. A set of dredges hangs from a steel beam with rubber wheels at each end, and this is towed along the seabed. The scalloper tows dredges on both sides and has a pair of derricks to lift them aboard.

Scallop dredges play havoc with the sides of wooden boats, so modern scallopers are built of steel. The smaller boats are constructed for the job. Many of them are dual-purpose scalloper/trawlers, which can work whatever fishery is more profitable. When they haul their dredges, they tip them individually with a gilson. Large scallopers are converted beam trawlers, which can tow sixteen dredges each side. Most have tipping systems that tip the whole set of dredges in one go, and conveyors for collecting the catch. Many of these large beamers came to work the West Country grounds from the Solway Firth. Modern boats migrate to grounds all around the British Isles; for example, in 1980 many West Country scallopers worked off Fishguard in Wales, and in 2012 they worked very profitably off the French coast in the Seine Bay.

Scalloping makes a big contribution to the West Country fishing economy, but there are also environmental concerns about the effects of scallop dredging. As a result, more controls have been imposed. The Scallop Fishing Order of 2012 restricts the number of dredges worked between 6 and 12 miles from land to eight a side, and scallopers are banned from Marine Conservation Zones altogether, which are designed to protect areas of special environments.

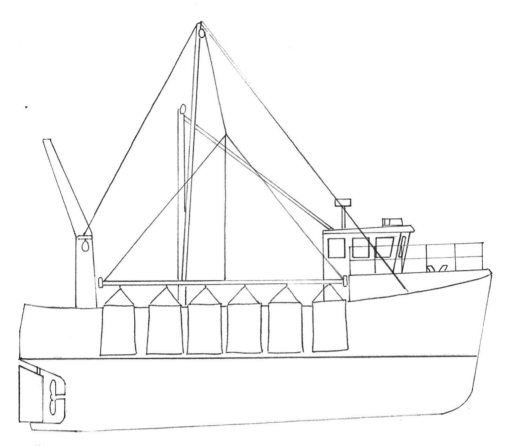

Scalloper.

Falmouth scallopers *Amethyst* FH 664 and *Rebecca* FH 665, 2014. They tow five dredges a side.

British Motor Fishing Vessels

SOUTH DEVON BEACH BOAT

The life of the South Devon beach fishermen in the early twentieth century was described by Stephen Reynolds, who worked a beach boat with Bob and Tom Woolley of Sidmouth. He detailed the hard physical slog of rowing 10 miles off to the fishing grounds, shooting herring nets, riding to them for several hours, hauling them, rowing back to shore if there was no wind for sailing, and landing and packing the catch. His first-hand experience encouraged him to become a highly effective Fishery Officer for the West Country, and to support the motorisation of the boats during the First World War.

The Britt engine, made at nearby Bridport, was popular. With the arrival of motors, boats were built a bit smaller – about 25 feet long instead of 28. The builders were Lavers & Dixons of Exmouth.

The Beer and Sidmouth boats were clinker-built on steamed frames, and had well-maintained varnished hulls. Beer was unique in Britain for working three-masted luggers – the last of them being broken up in 1918. The luggers went hand-lining for mackerel, towing up to four lines. The traditional withy crab pots were worked during the summer, while the herring season was from about October to January.

In 1935 Philip Oke recorded the Beer lugger *Little Jim* E 159, which was built in 1916 by Lavers of Exmouth for H. Bartlett. She had fourteen clinker planks a side on forty closely spaced steamed frames and was 23 feet long overall, with an 8-foot 6-inch beam and an inside depth of 4 feet 2 inches.

Beer is still a fishing beach, with a small fleet of fibreglass or carvel-built potters and netters. Its traditional varnished craft, used as tripper boats, have been refitted as Beer luggers, with a dipping lug foresail and a standing lug mizzen. The foresail sets well forward and hooks to an iron bumkin, which slopes down from the bow. The dipping lugsail must always be on the lee side of the mast, meaning it must be dipped to reset it each time the boat tacks. The Beer sailors have devised their own rapid flip-around method for this chore, and their weekly races are lively events. Many of these boats were built by the Harold Mears yard at Axmouth. From the 1980s, carvel building became more usual for fishing boats as they are easier to maintain.

Modern boats have the wheelhouse forward and are fitted with hydraulic haulers for their nets and pots. *Maverick* E 65, a recent Cygnus GM 26-foot fibreglass boat fitted out by Blue Water Boats of Plymouth for Beer, is strengthened for working off the beach, and has a reduced draught and stainless steel protection on her keel and bows. She has a JCB diesel engine and a Spencer Carter pot hauler and net hauler.

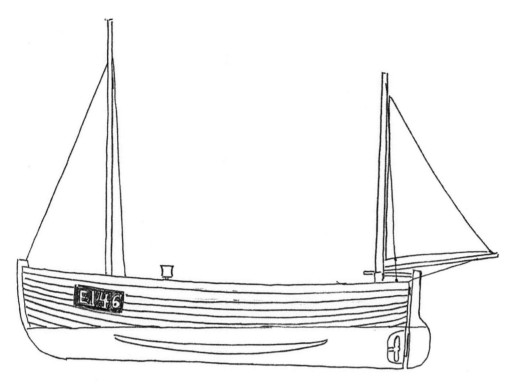

South Devon beach boat.

The varnished Beer beach boat *Sylvia* E 253 was built by Dixons of Exmouth in 1943 and rigged for trawling. She has a capstan for hauling her trawl, behind the foremast. Her wheelhouse is an interesting structure, as is her neighbour's, E 177.

British Motor Fishing Vessels

SOUTH DEVON CRABBER

In the post-Second World War years, the crabbing industry in South Devon developed. Browse Brothers of Paignton built up their business, which included catching, processing and marketing. In 1953 they experimented with canning crab; the following year they turned over to quick-freezing crab meat. They had their own fleet of boats, like the 41-foot *Skerry Bell,* built at Porthmellon, and the 44-foot *Torbay Pearl* DH 29 and *Torbay Queen* DH 38 – all of which were fitted with radio telephones. Their 44-foot boats had four crew and worked about 300 pots in strings of forty-five. Their dome-shaped pots had a heavy wire mesh base and a frame of wire and split hazel, which was covered with netting. The usual crab pot bait was ray or gurnard.

The typical South Devon crabber was built for the job, with a low freeboard, graceful lines and flared bows, the cabin and pot hauler forward, the wheelhouse aft, and completely unobstructed decks (unlike the bigger Cornish boats, which were full of hatches for working drift nets). The earlier crabbers had square sterns, but many of the later boats had round cruiser sterns, such as the *Torbay Queen* DH 38, which was built at Galmpton on the River Dart in 1959. She was 42 feet long with a 14-foot beam and a depth of 4.3 feet. Renamed *Francis*, she trawled from St Ives from 1964 to 1971, and then had a very long career as a netter and trawler at Mevagissey.

In 1964 Dixons of Exmouth built the 32-foot-long *Blossom Jill* PE 48 for Twynham Trawlers of Lymington. She was powered by a Gardner 54 hp diesel engine and was equipped with a hauler made by Mann of Exmouth, an Ajax 25 radio telephone and a Kelvin Hughes echo sounder. The following year they built the 32-foot *Southern Head* NN 99 for Skipper Peter Storey of Newhaven. 32 feet long with a 12-foot 6-inch beam and a 4-foot 6-inch draught, she was driven by a Perkins 115 hp engine and was fitted with a Decca Navigator, a Ferrograph echo sounder and a Sailor radio telephone. She worked up to 300 pots. In 1970, Dixons built the *Bolt Head Queen* SE 27 for Salcombe.

In 1974 the *Skerry Belle* DH 63 began her potting career off Start Point. Built by Dixons for a Kingsbridge owner, and powered by a Gardner diesel engine, she had a length of 39 feet 6 inches, a 14-foot beam and a 5-foot draught. Her hauler, built by Hubbard of Brixham, was fitted with a davit from Brixham Steel Construction for raising her pots to working level. This soon became standard for crabbers. Recently sold to Newlyn, the *Skerry Belle* is still going strong.

Several of these craft were built for trawling and spratting. For example, the *Heart of Oak* E 14 was built of larch planks on oak frames in 1978 by Dixons for West Bay Bridport. She was 32 feet long with an 11-foot beam and a draught of 4 feet, and was powered by a Lister 43 hp diesel. She was fitted with a Brixham trawl winch and a Ferrograph echo sounder.

In 2017 the 34-foot crabber *Lucky Seven* E 262, built by Dixons in 1979, was still working from Lulworth Cove in Dorset.

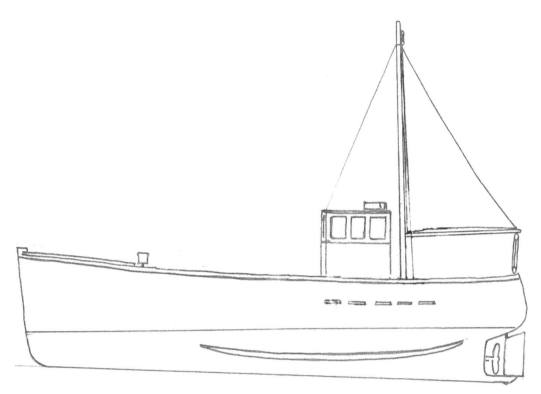

South Devon crabber.

The cruiser-sterned Salcombe crabber *Newbrook* SE 33 at Brixham in 1962. The 47-foot *Newbrook* was built in 1960 and was still at work from Dartmouth as DH 149 fifty years later, powered by a 130 hp engine.

British Motor Fishing Vessels

BIG DEVON CRABBER

The development of improved navigational aids, including the Decca Navigator and Radar, made it easier for the crabbers to explore grounds further offshore without the danger of losing their expensive gear. The excellent craft described on the previous page were really day boats, returning to their home port each evening. From the late 1960s a new generation of bigger craft were constructed with higher freeboard and live-aboard accommodation.

In 1966 G. Percy Mitchell of Portmellon completed the *Excel* DH 84 for Browse Brothers. The *Excel* was 49 feet 6 inches long with a 15-foot beam and a 6-foot draught. Planked with iroko on oak frames, she was driven by a Kelvin 112 hp engine, giving her a speed of 10 knots. Her wheelhouse equipment included a Ferrograph echo sounder and a Curlew radio telephone. Her cabin was fitted for four crew. A second *Excel* DH 17 was built at Appledore in 1971.

In 1969 the same company took the then unusual step of ordering a steel crabber from the Bank End shipyard at Bideford. The 45-foot *Sancris* DH 100 had a beam of 14 feet 9 inches and a draught of 5 feet 6 inches, with accommodation for a crew of three. She was also powered by a Kelvin 112 hp engine and was based at Dartmouth.

The 52-foot *K.M.B.* DH 81 arrived from her Danish builders in 1972. Skipper Kenneth Browse started work on the Hurd Deep grounds in the English Channel. She was a sister ship to the Appledore-built *Excel,* and the thirty-second vessel built for Mr Browse. She was built of oak on oak frames in only four months by the Nexo Skibs yard in Bornholm, Denmark, as British yards could not ensure a fast delivery. The *K.M.B.* fished 400 pots and began work from Poole in Dorset, later moving to Dartmouth. She was fitted with radar. Her engine was a Volvo Penta 185 hp and she had accommodation aft for five crew.

Two years later Hinks of Appledore built the slightly shorter *L.B.P.* for Logan Brothers and Skipper Tom Preston of Salcombe. The hull was designed by Gary Mitchell of Portmellon. She had a length of 48 feet with a 16-foot beam and a 6-foot 9-inch draught. There was a store forward, her Gardner engine was amidships, and there was a cabin for four aft.

In 1978 Skipper Ken Browse took over the 55-foot *Crusader* DH 71, which was also designed by Gary Mitchell and built by Hinks of Appledore. She was the biggest boat in the Browse fleet. In 1984 the Hinks yard began work on the transom-sterned 55-foor crabber *William Henry* from the same designer for Skipper Rick Mitchelmore of Dartmouth. She was fitted with a vivier seawater tank to keep her catch alive.

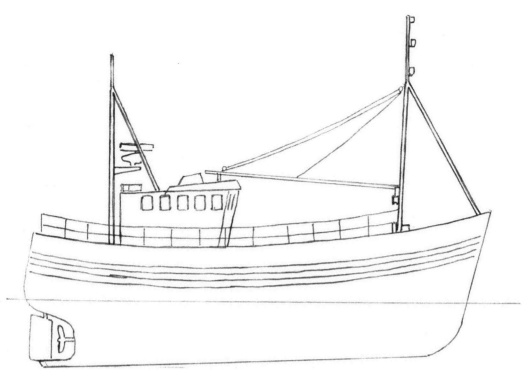

Big Devon crabber.

The steel crabber *Chris Tacha* PL 26, which was built in 2003. From the late 1980s, big crabbers were built of steel. Their decks are covered by shelter decks and they are fitted with pumped shellfish tanks (viviers). The early steel crabbers were known as super crabbers. They are capable of fishing all around the British Isles.

British Motor Fishing Vessels

FRENCH CRABBER

The first Camaret crabber, the cutter-rigged *Aventurier*, came to fish off the Isles of Scilly in 1902. The French crabber had been invented only three years before to fish the deep water Rochbonne ground, south of Brittany. What was new about them was that they were safe, decked boats, fitted with shellfish tanks, or viviers, to keep their catch of crawfish alive. The *Aventurier*'s voyage was a success and she was soon followed by a whole fleet of crabbers from Camaret, Loguivy, near Paimpol and Le Conquet. The Cornish and Scillonian grounds were a Klondike for the Bretons and they worked them successfully until the extending of British territorial waters in 1964 excluded them from most of their grounds.

The crabbers had changed since the little gaff cutters of 1902. By the mid-twentieth century, they were powerful motor boats, all fitted with echo sounders for finding their fishing grounds and radio telephones. Many of them had radio direction finding (DF) loops to help them navigate. The extending of British territorial waters and the depletion of the crawfish stocks clipped their wings, and from the 1960s many Camaret and Audierne crabbers were for sale.

Meanwhile, the shellfish industry was developing in the Channel Isles and West Country. The Breton crabbers were regarded as ideal for the job, being fine seaworthy boats that could carry large numbers of pots and keep their catch in prime condition in their viviers. They were snapped up by fishermen in the Channel Isles, Dorset, Devon and Cornwall.

Perhaps the first was Raphael Ansquer's Audierne crabber *Nicole* AU 2422, which was sold to Barlow Richards and registered in Penzance, but was sadly lost in 1965. Fortunately, her crew were safe. In 1966 John Burt of Newquay bought the *Bacchus* CM 3027 from Camaret and registered her as PW (Padstow) 251. John and his crew continued to work her in much the same way as her Breton crew. In 1968 Pat Crockford of Falmouth bought the *Etoile du Marin* AU 2293 from Audierne, registering her as FH 4, and Jim Richards bought the *Francois and Jeanette* and registered her as PZ 22. The Audierne crabber *Keristum* AU 2540 was sold to Guernsey in 1988. In 1994 she was bought by Rick Mitchelmore and was registered as DH 181 at Dartmouth. Sold to Scotland in 2002, she was registered as BCK 77 and then SY 76. The *Le Cap* AU 2479 was sold to Jersey as J 91 in 1981. In 1991 she was bought by Harvey & Co. of Newlyn and was renamed *Rachel Harvey*.

The ex-Breton crabbers were adapted for their new role. A light tripod took the place of the tall sailing foremast, while the mizzen mast was replaced by a structure to carry their radars and aerials. The wooden wheelhouse and galley were often renewed in steel. Over the galley was the podium – a wide shelf, with high rails, to carry up to four layers of pots. The whole boat was surrounded by rails to contain the vast numbers of pots needed for crabbing. There was often a shelf over the stern, known as the cat catcher, to carry the bongos (large plastic drums) used for handling the catch. Forward, on the starboard side, was the grooved slave hauler, or vee wheel, for hauling the gear, and above it was the pot davit for raising the crab pots to a convenient working height. This got rid of the chore of hauling them up. Many of these boats, properly maintained, fished on for half a century.

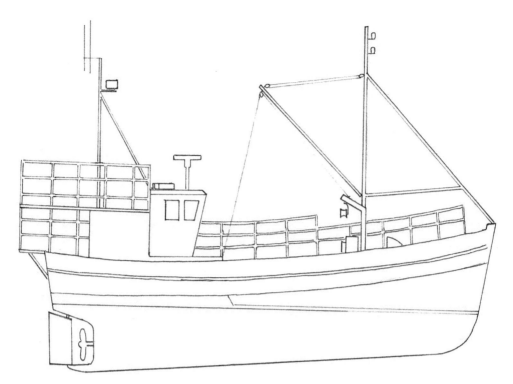

French crabber.

Skipper Rowse's ex-Breton crabber *Stereden va Bro* TO 5 working her gear. Her name is Breton for 'Star of the Land'. (Photograph Glyn Richards)

British Motor Fishing Vessels

BEAM TRAWLER

The beam trawl, a stocking-shaped net kept open by a long pole known as the beam, is an ancient invention, and was used by sailing smacks until the 1930s. Twin beaming (towing two beam trawls from derricks on either side of the trawler) is a method used for shrimp trawling in Germany, Holland and Belgium. By the 1960s twin beaming was being used very profitably by the Dutch for catching white fish; in the spring of 1971, over thirty Dutch beamers were based on Fleetwood, working the Morecambe Bay sole fishery. These modern trawls have a steel beam with a frame called a trawl head at each end to raise it off the ground. On the bottom of the net is the chain mat – a grid of chains that prevents stones from entering the trawl. When towing her trawls, the beamer lowers her derricks down on either side. When the nets are hauled, they are not brought aboard; rather, as the trawls are hauled up on each side, only the cod ends of the nets are brought aboard and emptied. The beam trawl is efficient at catching high-value flat fish, but the beamer guzzles fuel to tow this heavy gear along the seabed. A typical Dutch beamer built in 1973 had a length of 31.80 metres, a beam of 7.50 metres, a hold depth of 4.10 metres and a 1,200 hp engine.

The Brixham fishermen were pioneer trawlermen in the days of sail. Their smacks discovered new fishing grounds all around Britain and contributed to the development of Ramsgate, Boston, Scarborough, Grimsby, Hull and Milford Haven as fishing ports. During the Second World War a large fleet of modern Belgian trawlers fished from Brixham. In the post-war years, the port developed with the creation of its own fishing co-operative. Several Belgian trawlers were bought. In 1967 the 75-foot modern Belgian beamer *Duc in Altum* N 800 from Nieuwpoort came to join the Brixham Trawler Race and evoked much interest. In 1969 brothers Tony and Quentin Rae bought the 57-foot beamers *Sara Lena* BM 30 and *Scaldis* from Holland. Both were powered by 240 hp Mercedes engines. They were immensely successful and were followed by other vessels. In 1970, the 68-foot *Catharina Kes* BM 2 was bought by Brixham owners Les Cunningham and Lukas Heller. Powered by a Kromhout Stork 190 hp engine, and equipped with a Decca 203 Radar and Track Plotter and a VHF radio, she was built in Urk in 1964. It was discovered that her 8-metre-long beams were too big for fishing in deep water and they were reduced to 6 metres.

At the time, the *Catharina Kes* was the biggest beamer in the UK, but she was soon overtaken as many large vessels were bought during the 1970s for Brixham, Newlyn, Lowestoft, Fleetwood, Grimsby, Hull and Portsmouth from Holland, which continually updated its beam cutter fleet. Other Brixham beamers were the *Catear* BM 282, which was bought in 1982, and the *Lloyd Tyler* BM 188 and *Jacomina* BM 208, which were bought in 1988. Lowestoft's fleet included the *St Matthew* LT 60, *St Martin* LT 62 and *St John* LT 88 for the Colne Group, the *Korenbloom* LT 535 for Dowsett Trawlers and the *Broadholm Queen* LT 344 for Talisman. Newlyn acquired a large fleet, including Stevensons' *Aaltje Adriaantje* PZ 198, *Algrie* PZ 199, *Cornishman* PZ 512 and *St Georges*, Nowell's *Elizabeth N* PZ 100 and *Louisa N* PZ 101, and Corin's *Sapphire* PZ 66. Irish-owned beamers included the *William Joseph* WD 182, which was built in 1972, the *Avontuur* WD 148, from 1973, and the *Marie Jacob* D 141 and *Kathleen K* WD 29, which were both from 1971. Several conventional side trawlers were converted to beamers in Holland, Belgium and the UK. The heavy beam gear severely tests trawlers' stability and, sadly, several have been lost.

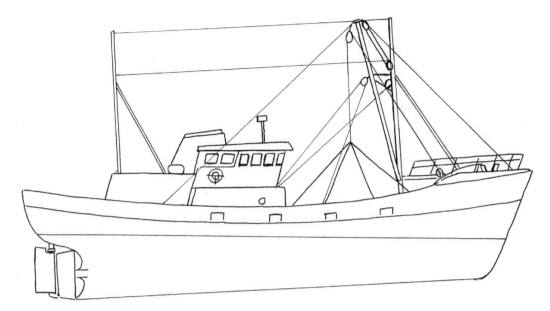

Beam trawler.

The beam trawler *Louisa* N PZ 101 in 2014.

British Motor Fishing Vessels

HASTINGS BEACH BOAT

Hastings has no harbour and its boats are launched and hauled up on its beach, which is called the Stade. Until the nineteenth century, Hastings sent big boats to the North Sea for the herring fishery and to Cornwall for the mackerel season. By the early twentieth century, there were two kinds of clinker-built boats: decked luggers of about 30 feet in length and smaller open boats called punts. The Hastings boats were finding it hard to compete with the big Rye and Ramsgate smacks that often landed locally. There was mackerel drifting from May until August, then trawling until October when the herring arrived, followed, after Christmas, by more trawling, until the next mackerel season.

The first boat to be equipped with an engine was the *Boy Leslie* RX 23 in 1914, and she was followed by the *Albert Edward* RX 16 and *Surprise* RX 130. Soon all the big boats were motorised, usually with 13 hp and 6 hp Kelvin petrol/paraffin engines. Most boats had a 6 hp and a 13 hp engine with, inconveniently, a propeller on each side. Since they were underpowered, tall masts, bowsprits and sails were retained. There was a boom in fishing during the First World War, followed by two decades of depression during the 1920s and '30s. The drift net fisheries for mackerel and herring were largely given up and the big boats trawled all year round. Motor capstans were fitted for hauling the trawl and beam trawls were replaced by otter trawls with trawl boards in the mid-1930s. The picturesque horse-powered capstans were replaced by motor winches for hauling up the boats. Three boats were mined during the Second World War, two of which were tragically lost with all hands. Fishing was dangerous but profitable. Since night fishing was forbidden, there was no drift netting.

In the post-war years there was a spirit of optimism. The White Fish Authority's Grant & Loan scheme helped to build a whole fleet of fine new boats for Hastings at boatyards in Newhaven, Rye and Whitstable in the late 1950s, all of which were powered by powerful diesel engines and fitted with wheelhouses.

Skipper Brian Stent replaced his *Valiant* with the *St Richard* RX 60 in 1969. She was built by Tankerton Bay Marine of Whitstable to a design by Christopher Cox. The *St Richard* was 33 feet long with a beam of 12.5 feet and a 4-foot draught. Of traditional clinker build, with elm planks on oak frames, she was powered by a Gardner 70 hp diesel with a 2:1 reduction gear. In the late twentieth century, trammel nets largely replaced trawling and brought prosperity to Hastings. The last traditional big boat was the *Our Pam and Peter* RX 58, which was built for Skipper Denis Barton by H. J. Phillips of Rye in 1980. The first steel boat arrived in the late 1980s and three of them now work from the Stade.

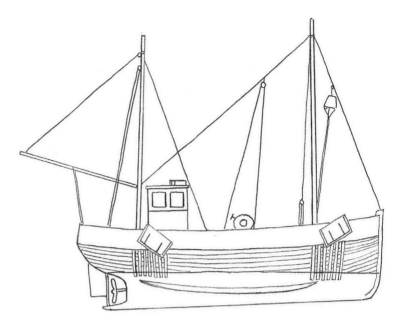

Hastings beach boat.

J.E.S.E. RX 322 fished from Hastings from 1948 to 1969, when she was sold to Bognor. She was later fitted with a bigger wheelhouse. Her trawl can be seen hanging up to dry. When artificial fibre courlene trawls arrived in about 1960, they no longer needed drying. A pair of oars hang over the side and her rudder is raised to save it from damage when beaching. On the beach are the wooden troes that the boats are launched over.

British Motor Fishing Vessels

HASTINGS PUNT

The smaller class of open boats at Hastings were the punts. In the days of sail, they were rigged with lugsails on both masts, sometimes with a sprit mizzen sail. Like the big boats, they were fitted with dagger boards to help them sail. For much of the early twentieth century, when the local industry was depressed, the cheaper, more economical punts managed to scratch a living by adapting to a range of inshore fisheries: drifting for sprats, trammel netting for white fish, mackerel drifting, herring drifting, shrimp trawling and long-lining for dogfish. The sailing punts were small, ranging from 16 feet to 20 feet in length.

Engines were fitted in the luggers during the First World War, but the punts, with their more modest earnings, were not motorised until the 1920s. In the years after the Second World War, the high cost of nets reduced earnings from trammels, but there were some profitable sprat drifting seasons. In the late 1950s trammel netting for soles and plaice took off. As small diesel engines became available, bigger punts were built, reaching up to 25 feet long. Hastings trammel nets were 50 yards long and under 2 yards deep, and were rigged in fleets of ten. Up to a dozen fleets were worked.

Traditionally, trammels were fished in the spring and summer, but as they were profitable they were soon worked all the year round. Such was their success that the big boats followed their lead and gave up trawling for netting. In the 1960s, herring drifting was an important season for the punts.

From the 1970s many more punts were built and they became an important part of the Hastings fleet. They were traditionally built with clinker planks and either lute or elliptical sterns. Sussex is the only place in Britain where the ancient lute stern is still seen. It is a square transom stern with the upper part curving over the rudder to prevent waves breaking into the boat. The introduction of increasingly powerful hydraulic net haulers and artificial fibre nets were a boost to the punt fishermen and enabled them to work much more gear and increase their earnings. In 2014 many boats fitted net stackers which replaced the time consuming chore of overhauling nets manually on to the beach at the end of each trip.

In 2017 there were fourteen boats working from the Stade, each nudged into the sea by a caterpillar tractor which hauls them up again at the end of each trip. Wooden punts continued to be built after the last traditional wooden big boat. The modern Hastings boats are nearly all built of fibreglass, including three large catamarans. Only two traditional wooden boats appeared to be still at work in 2017 with others laid up on the beach. The modern Hastings fleet has diversified. Although gill netting and trammelling are the main business of the beach, some boats have worked drift nets, pots, whelk pots and scallop dredges. The port is believed to be unique in having the biggest beach fishing fleet in Europe.

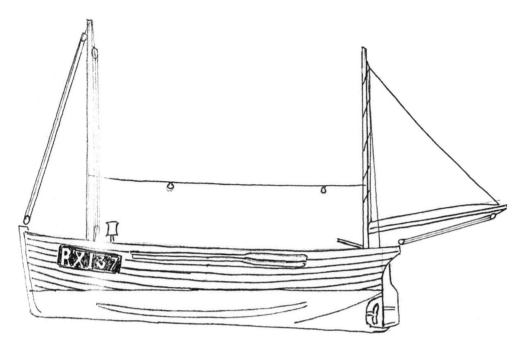

Hastings punt.

Lute-sterned Hastings punt *Four Sons* RX 137 in 1968. She has a capstan for hauling her buoy ropes but no hydraulic hauler for working her trammel nets. These arrived in the 1970s. Her nets are worked by hand.

British Motor Fishing Vessels

THAMES ESTUARY AND ESSEX COAST

The Thames Estuary and Essex coast were the home of the sailing smack and the square-sterned bawley. The smack, based on the Colne, Blackwater and Crouch rivers, was mainly an oyster dredger, but also went trawling, spratting and shrimping. The Bawley, primarily a shrimp trawler, fitted with a boiler to boil her catch, belonged to the Thames and was also based at Rochester, Whitstable, Leigh and Harwich. Motors were fitted into these boats from before 1914, and motor bawleys were built from the 1930s. Many smacks and bawleys had long careers with auxiliary motors.

Spratting took off after the Second World War, and canneries were built at Whitstable and Wivenhoe. The Leggat brothers introduced midwater pair trawling at Whitstable and built the smartly varnished sister ships *Romulus* F6, *Remus* F7 and *Faustulus* F21 in 1954. These boats later went spratting from Wells-next-the-Sea. North Sea Canners of Wivenhoe had the 48-foot trawlers *Essex Girl* CK 54 and *Fisher Girl* CK 59 built to extend the winter sprat season until May. Pair trawling replaced the ancient method of stow boating, which was used by the sailing smacks. The stow net was a large stocking-shaped net, which was kept open at the mouth by two hefty wooden spars called baulks, and was anchored in a likely channel for a shoal of sprat to swim into.

The traditional craft were not powerful enough for midwater trawling and several second-hand boats were bought. At West Mersea, Skipper Peter French bought the 48-foot Scots ring netter *Clauran* CK 100 in 1963 for single boat midwatering. In 1969 she was replaced by the wooden stern trawler *Providence* CK 1, which was built by R. J. Prior of Burnham. The *Providence* had a length of 48 feet 10 inches, with a 15-foot 3-inch beam and a Kelvin 112 hp diesel.

In 1966 Sutton & Wiggins of Great Wakering built the *Paul Peter*, designed by F. Parsons of Leigh for Peter Gilson of Southend. She was 41 feet long with a 12-foot 6-inch beam and her draught was only 3 feet 4 inches. Her Gardner 84 hp engine gave her speed of 9 knots. The wheelhouse and cabin were forward, followed by the fishroom and the engine was aft. Her trawl winch was belt-driven from the fore end of the engine. She was intended for bottom trawling and midwatering for sprats, and was fitted with an Ajax radio telephone and Kelvin Hughes echo sounder. She was fitted with one trawl gallows forward on the port side. The other trawl warp went through a fairlead in the middle of the transom stern. A sister ship, *Three Sons*, was to be built for Laurence Gilson.

In 1967 Anderson Rigden & Perkins Ltd of Whitstable built two sister ships for local skippers, the *Tykela* F26 for Victor Davis and the *Kordella* for E. Hoy. They were 43 feet long with a 12-foot 6-inch beam and a shallow draught of 3 feet 4 inches. Built of iroko on oak or rock elm timbers, they were powered by Gardner 84 hp diesels. Both were fitted with autopilots, echo sounders and Ajax radio telephones. Their wheelhouses were forward. In 1971, Vic Davies had the 45-foot stern trawler *Ticino* F 44 built at the same yard. Her engine was a Gardner 110 hp.

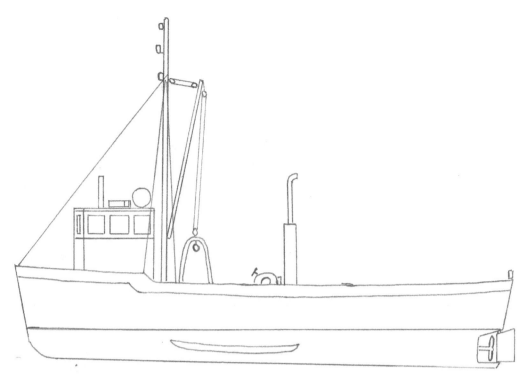

Whitstable trawler.

Genesta LO 49 at Leigh in 1975.

British Motor Fishing Vessels

LEIGH COCKLER

The earliest nineteenth-century sailing cocklers were ex-Naval galleys fitted with a gaff sloop rig. Soon, shallow-draught cockle boats were built for the job. They were among the few British working boats to be fitted with centre boards (the others were Suffolk punts, Hastings luggers and Cornish covers and gigs). Cocklers fished by grounding on the Maplin Sands, where their crews raked up and loaded up to 2 tonnes of cockles before the tide returned and floated the boat. While bawleys boiled their shrimps aboard, cockles were taken back to the waterside sheds at Leigh to be boiled there. When motors were fitted, it was discovered that the propeller could be used to wash the sand off likely cockle beds before the boat grounded.

The motor cockler *Resolute* was built for C. & W. Osborne of Leigh by Hayward of Southend in 1927. She had a raised foredeck to give more room in her cabin and a slightly flared bow to keep her dry. The *Resolute* was strongly built with sawn oak frames, pitch pine planks and a yellow deal deck. She was 36 feet long with an 11-foot 4-inch beam, a moulded depth of about 4 feet 2 inches and a draught of only 1 foot 6 inches. Though a motor boat powered by a 15 to 18 hp Kelvin petrol/paraffin engine, she was also fitted with a 6-foot centre board to help her sail. There was accommodation for six in her fore cabin; next came the hold and the engine room was aft. Her mast was in a tabernacle and she carried a boomless gaff mainsail, staysail and jib.

Raking cockles was a labour-intensive business. In 1967, the suction dredge was first tried. This involved the boat steaming over the sand in between 5 and 15 feet of water and then lowering the dredge over the side. A water jet from the dredge dislodges the cockles, which are sucked up through a pipe. They would go through a rotating screen, which separated out undersized cockles, mud and sand. The maximum permitted 3 tonnes could be dredged in three hours, with the boats working two days a week for a limited season to conserve the stock. Hand cocklers, on the other hand, worked all year round.

The cockle dredge and its engine to power the pump and associated gear are a big load for a traditional wooden boat. The last wooden cockler built was the *Reminder* LO 38. Since the 1980s, flat-bottomed steel boats have been built for suction dredging. The steel cockler *Renown* LO 88 was built for the Osborne family of Leigh in 1991 by Newbury Engineering of Newhaven, and in 2008 the family built the cockler *Mary Amelia* LO 86. Like *Renown*, *Mary Amelia* is also a traditional name, which previously belonged to a sailing cockler.

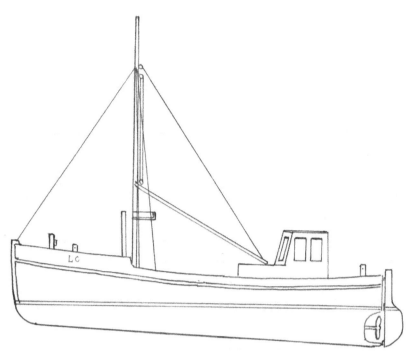

Leigh cockler.

Leigh cockler *Renown* LO 83 was built by Haywards of Southend in 1928. She had a raised foredeck to give some extra headroom in her cabin. Though a motor boat, she still had a full rig. Sadly, the *Renown* was lost in 1940 at Dunkirk with her four crew. She was one of six Leigh cocklers that sailed to Dunkirk to help rescue the Army.

British Motor Fishing Vessels

STEEL COCKLER/SHRIMPER

The modern shallow-draught steel boats that fish from Leigh on the Thames Estuary and King's Lynn and Boston on the Wash are versatile vessels that can adapt to the current season in areas where the ecology is fragile and the fisheries are highly regulated. This situation has been complicated by the building of wind farms in the Wash in an important mussel-fishing area. Today's steel boats are a far cry from the venerable motorised smacks that worked for much of the twentieth century.

Cockles live in the sand for two to four years and filter feed on plankton. In the Wash, boats from Boston and King's Lynn began prop washing the sand off the cockle beds in the 1970s. The hydraulic cockle dredge came into use in 1986 and by the late 1980s over thirty steel boats were dredging up to 10 tonnes a trip. Most of the catch was exported to Holland and Spain. Concerns about the ecology of the area and its resistance to mechanical fishing methods resulted in the Wash Fishing Order of 1992. It was realised that the hydraulic dredge might not be the best way of fishing. Undersized cockles did not always survive screening. With boats allowed to dredge up to 8 tonnes a day, a whole season's total allowable catch could be caught in just a few weeks. In the early 2000s the dredged catch was reduced to 4 tonnes per boat per trip. A further step was to permit hand raking only, with a total catch of 2 tonnes daily, which has been enforced since 2009. The result has been healthy cockle stocks.

The boats prop wash a likely cockle bed before grounding on the sand, where their crew rakes up to 2 tonnes. The cockles are raked into nets, which are tipped into baskets. These are tipped into 1-ton bags that are then hoisted aboard by the boat's hydraulic crane. In 2015 up to 2,079 tonnes could be caught. By contrast with the large fleet in the Wash, a very small fleet of only fourteen licensed boats continued to dredge very profitably on the Thames Estuary in 2016.

A fleet of sixty boats is based at Boston and King's Lynn. It can adapt to fish cockles, mussels, brown shrimps and pink shrimps when permitted. There has also been a sprat fishery, but in 2015 this was limited by lack of quota. Additionally, whelking is increasing in importance, and mussels are also dredged. A mussel dredge is a mesh bag with a blade at the entrance, which is towed along the seabed by the dredger. Shrimps are caught with twin-beam trawls. All the boats are fitted with a goalpost mast and twin derricks to handle them. In 2016 the see wing beam trawl was tested for shrimping. This trawl has an aerofoil-shaped beam, which is intended to reduce the net's impact on the seabed. The shrimp fishery has been assessed by the Marine Stewardship Council (MSC).

Britain's three major shrimping areas were the Wash, Morecambe Bay and the Thames Estuary. The Wash is now the most productive fishery.

A typical modern boat is 14 metres long with a hefty 6-metre beam and a draught of only 1.5 metres, is powered by a 300 hp engine with a 170 hp engine to drive the pump, and is fitted with a bow thruster, autopilot, radar and plotter, echo sounder, hydraulic crane and up to six winches for handling her variety of gear. Recent boats have been built by Newbury Engineering of Newhaven and Harris & Garrod of Grimsby. They are chunky shallow-draught vessels that look like toy boats from a distance. Thanks to the careful control of their work, however, it looks as though their future is secure.

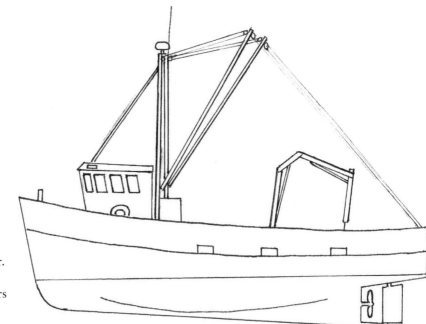

Right: Steel cockler/shrimper.

Below: Shrimpers at Fisher Fleet, King's Lynn.

British Motor Fishing Vessels

SUFFOLK BEACH BOAT

The traditional 15 to 20-foot-long clinker-built beach boats of Suffolk sailed from Bawdsey, Shingle Street, Aldeburgh, Thorpeness, Dunwich, Southwold, Kessingland and Pakefield. Some of them have survived into the present century and there are plans to restore and maintain several of them. Built of oak or larch planks on steamed oak frames and sawn oak floors, the sailing punts were rigged with a dipping lug mainsail on the foremast and a standing lug mizzen sail, which sheeted to a long outrigger on the starboard side of the rudder. They were flat-bottomed and some were fitted with centre boards to help them sail, but most relied on bags of shingle ballast that could be tipped out if the boat was laden down with a good shot of fish. In common with many British clinker-built boats, they were painted white with a coloured top strake. They thrived in the twentieth century with up to twenty working from Aldeburgh in the early 1980s. Motors were fitted from the First World War and motor boats were built from the 1920s. Small diesel engines did not become available until the 1970s; until then, petrol engines were used and sails were sometimes retained.

The Suffolk beach boats trawled for soles, plaice and ray during the summer, took holidaymakers out for trips, shot long lines for cod, and went drifting for high-quality longshore herring in the autumn and for sprats in the winter. Some trawled for shrimps or worked lobster pots. The sprat season was the big earner, but unmeshing sprats from drift nets onto the beach was a fiddly business.

In more recent times, many were worked by part-time fishermen. Problems with marketing small and unreliable quantities of fish were solved by selling directly to the public from their fishermen's sheds. Some of the latest boats were very well equipped, even being fitted with miniature trawl winches complete with wires and warping drums.

In 1966 Frank Knights, boatbuilders of Woodbridge, built the beach boat *C & R* IH 101 for Billy Burrel of Aldeburgh. Named for her skipper's children, Caroline and Richard, she was 19 feet 3 inches long with a massive 8-foot 2-inch beam and a 2-foot draught. The *C & R* was copper-fastened and planked with ½-inch larch on Canadian rock elm frames spaced at 5-inch centres. Her engine was a 13 hp air-cooled Lister diesel. She was fitted with a Jabsco pump, belt-driven from the engine, and a Whale hand pump. Her forward capstan was driven by a belt from the engine. There were three iron-shod oak bilge keels on either side to protect her when landing. In 1973, Frank Knights built the *Four Daughters* IH 158 for Sid Strowger of Aldeburgh.

In 1968 the Felixstowe Ferry Boatyard built two beach boats, the *Eileen R* IH 111 and the *Moonraker*, intended for potting and long-lining. The *Eileen R* was 18 feet long with a beam of 7.5 feet, and was built of larch planks on steamed oak frames. Her 9 hp Lister air-cooled diesel gave her a speed of 7 knots. The capstan was driven off the engine. Although having a reputation for being noisy, air-cooled engines were suitable for beach boats. With water cooling, there was always the danger of sand entering the water inlet. An alternative was a keel cooling system, but this was vulnerable in boats that were launched and hauled up each day.

The modern Suffolk inshore fleet of GRP-hulled boats works from Aldeburgh, Sizewell, Orford, Felixstowe Ferry and Southwold, lining and netting for sole, bass and cod.

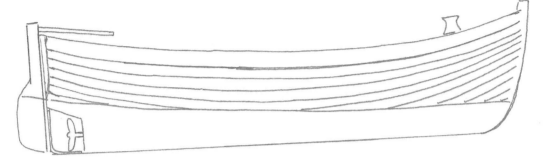

Suffolk beach boat.

November sprats at Aldeburgh. (Photograph Pam Oakley, Hendon WI)

British Motor Fishing Vessels

NORTH NORFOLK CRABBER

The roomy clinker-built North Norfolk crab boats of Mundesley, East and West Runton, Cromer, Weybourne and Sheringham were constructed with pointed sterns to cope with the sea conditions experienced when launching from these beaches. The design of the boats is older than the crab pots that were introduced from the north-east of England in 1863. The highly efficient local pot was made from four hazel hoops stuck into a wooden base and covered with netting. Before crab pots, the primitive hoop net was used. The crab boats were rigged with a high-peaked dipping lug sail, which hooked onto the bow and was sheeted right aft. There were six oar holes called orrocks cut into the top strake and a long rudder that extended below the keel. When the boat was beached she was heeled towards the shore, so the waves of an incoming tide would knock her up the beach. Once all loose gear had been removed, oars could be inserted through the orrock holes, and these were used to carry the boat up the beach. A typical sailing crabber was about 19 feet long with 7-foot beam and a depth of 3 feet.

Petrol or petrol/paraffin motors were fitted in the crabbers in the 1920s and '30s, but sails were often retained. Motor crab boats were fitted with only four orrocks. Oars were still needed to keep the boat straight when she was being launched. The wide stern post with an arch to protect the propeller was introduced by boatbuilders Emery of Sheringham, who built crabbers until 1981. Then fishermen turned to Maycraft of Potter Heigham on the Norfolk Broads, who built the last wooden crabber, the 22-foot *Valerie Teresa*, in 1989. Worfolks of King's Lynn also built crabbers.

The sailing boats had no gunwales, to keep them light, but they have been fitted since the arrival of Duerr hydraulic capstans for hauling the shanks of pots in the 1960s. Other British fishermen had used haulers since the 1930s.

When boats were built for engines, they were made a plank higher and were sometimes fitted with a washboard forward to keep them dry. They were also bigger. The last boats were up to 22 feet long with an 8-foot beam and were fitted with three thwarts. They typically had eleven larch planks on each side and the keel, stem, stern and the steamed timbers were of oak. Sometimes, the top planks were also of oak, or elm.

The first fibreglass crab boat, the *Paternoster* YH 63, was built by Stratton Long of Blakeney in 1974 for Richard Davies of Cromer. Based on the wooden *Charles Perkins*, and powered by a Thornycroft 44 hp engine, she was 22 feet long with an 8-foot beam and a 1-foot 10-inch draught. Motor boats were too heavy to be carried so they were pushed over skeets (rollers enclosed in wooden boxes). Fishermen worked together to launch the boats and winches were used to haul them up. Since the 1960s, Norfolk fishermen have had their own tractors and trailers for launching and recovering the crab boats.

Their main business was potting, but they also went herring drifting in the autumn and long-lining for cod and ray over the winter. The lines were worked from tin baths and were baited with mussels, whelks or herring from Lowestoft or Yarmouth.

The traditional wooden crab boats worked by two or three crew fished on until the mid-1990s, when they were replaced by high-speed outboard-powered GRP skiffs built by local company Tactile Boats of Edingthorpe. These 19-foot skiffs are square-sterned boats with a speed of 12 to 14 knots. Like other crabbers they are fitted with radios, GPS, echo sounders and slave haulers (which are forward instead of aft, as in the traditional boats). As an answer to the lack of experienced crews, the skiffs are worked single-handed and fish up to 250 pots in shanks of ten to fifteen. In 2015 there were fourteen boats working from Cromer – five at Sheringham and twelve at Wells harbour. There remained only one traditional wooden crab boat at work, the *Mary Ann* YH 213, which was built by Maycraft in 1973. Her skipper, John Jonas, has adapted her to work single-handed, like the skiffs. But the traditional crab boats are an important part of the local story and, as with the punts of the Suffolk beaches, several have been restored for leisure use by Rescue Wooden Boats of Stiffkey.

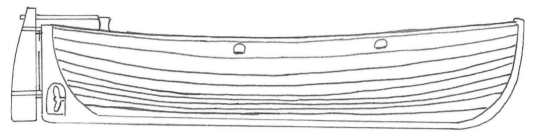

North Norfolk crabber.

Crab boats *Reaper*, *John Robert* and *Peggy* at Cromer in 1934.

British Motor Fishing Vessels

WHELKER

Whelking was an important fishery at Cromer and Sheringham. Several of their fishing families moved to Wells-next-the-Sea, which has a proper harbour, albeit a tidal one with a sand bar at its entrance. Since some of the whelk grounds were found 30 or more miles offshore, a bigger class of clinker-built motor whelker was developed during the 1930s. Typical of the class was the *Knot*, which was 26 feet long with a 10.5-foot beam. The 28-foot *Bessie* LN 16, built by Johnson in the mid-1930s, had a long career whelking from Wells. Among the biggest were the 30-foot *Salford* and *William Edward*, which were built for Cyril Grimes in 1949. Although heftier than the crab boats, these were still open boats.

The traditional whelk pot was dome-shaped, and was made of an iron frame bound round with rope. They were dipped in tar to preserve them. In the 1950s they were worked in shanks of twenty with anchors and dans at each end in the same way as crab pots. Unusually, they were shot against the tide. By the 1960s, numbers had increased to six shanks of thirty-six, and they were usually baited with salt herring from Yarmouth. The whelks were boiled in coppers ashore.

Several ex-Royal National Lifeboat Institution lifeboats have been converted to fishing boats, but Wells was unusual in having quite a fleet of them, which were bought in the early 1950s. Three of them, the *Anne* LN 175 (ex-Skegness RNLB *Anne Allen*), *Spero II* LN 162 (ex-RNLB *Howard D,* which served at St Helier, Flamborough and Arbroath) and the *Elizabeth* (ex-Seaham RNLB *Elizabeth Wills Allen*) were owned by the Cox family. Alan Cooper had the *Anne Isabell* (ex-Kilmore Quay RNLB *Anne Isabella Pyemont)*. These were all ex-Liverpool-class lifeboats, 35 feet long with a 10-foot beam and a 2-foot 3-inch draught. Developed during the 1930s, the Liverpool boats were originally powered by 35 hp RNLI petrol engines. Skipper Chic Smith had the Watson-class ex-lifeboat *Amethyst* LN 68. Although these boats had no shelter for their crews, just a curved canopy to protect the engine, they were decked boats, so were an improvement on the open whelkers. They were fitted with a capstan aft for hauling their shanks of pots and a little mast in front of the canopy to carry their lights.

In the 1970s the lifeboats were replaced by purpose-built decked boats fitted with wheelhouses aft. In 1977 the *Spero II* LN 162 was replaced by the new *Four Brothers* LN 101. Skipper Frary built the *Alison Christine*, which was fitted with radar and modern electronics. Fishermen now find their gear with GPS. In 1987 Alan Cooper had the 32-foot *Ma Freen* built by Goodchild at Yarmouth. By the 1980s, whelkers were working up to 40 miles offshore, shooting six shanks of forty pots and landing over a ton a trip. EEC rules banned the local boiling of whelks, which were from then sent away for processing.

In the 1940s the blacksmith charged 4*s* 6*d* for the metal frame of the pot. By the 1980s, this had risen to £11 for the frame and £6 for the iron base. The traditional iron-framed pots have been replaced by ones made from 6-gallon plastic containers perforated with holes to allow them to drain when they are hauled. The top, or sometimes the side of the container is cut off and replaced with a piece of netting with a hole in the middle. This enables the whelks to get in, but prevents escape. The base of the container has a concrete weight to keep it upright on the seabed and a strop to take the bait. Fishermen often buy commercially produced versions of these pots, which are more effective than the traditional model and keep out crabs and other predators.

Like other fisheries, whelking has seen times of boom and bust. Recently, several large decked boats have been adapted for whelking. These craft can fish anywhere, using large amounts of gear.

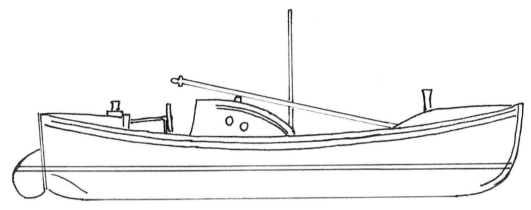

Whelker.

Whelker *Anne* LN 175, the ex-Liverpool-class RNLI lifeboat *Anne Allen*. (Image courtesy of Tony Denton, Lifeboat Enthusiasts Society)

British Motor Fishing Vessels

LOWESTOFT MOTOR SMACK

In the early twentieth century, Lowestoft had a large fleet of trawling smacks, many of which sailed to Padstow in the spring to fish for soles on the local Trevose grounds. In April 1911 there were ninety-three Lowestoft smacks at Padstow, together with twenty-four from Brixham and twenty-one from Ramsgate. During the First World War these vessels were decimated by German submarines, with eleven being sunk on one day (12 March 1917) off Trevose. With the coming of peace, many owners ordered replacement smacks from local yards, and from the highly reputed Rye boatbuilders, including the *Master Hand* LT 1203 in 1920, the *Helping Hand* LT 1239 in 1921, and the *Holkar* LT 18 and *Rosary* LT 73 in 1924.

Although Lowestoft retained a large fleet of sailing smacks until the 1930s, eighteen were converted to motor power. Their low-powered engines were initially auxiliaries and the vessels retained a full sail plan, sometimes with the addition of a small wheelhouse aft, and continued to work beam trawls using their steam capstans. However, many of these vessels were soon refitted as fully powered trawlers, driven by more powerful Crossley, Deutz, Allen or Petter engines. They had trawl winches in front of the wheelhouse and gallows on both sides. Their sails were removed, except for a small mizzen. Many of these fully converted smacks, including the *Try On* LT 176, *Holkar* LT 18, *Rosary* LT 173 and *Purple Heather* LT 249, were fitted with a wheelhouse, engine casing and galley similar to those on steam drifters.

The main owners of the motor smacks were W. H. Podd, Diesel Trawlers and Inshore Trawlers. W. H. Podd's vessels included the Lowestoft-built *Gleam* LT 293, which was launched in 1922, the *Pathway* LT 397, from 1915, and the *Pilot Jack* LT 1212 and *Flag Jack* LT 1224, which were built at Oulton Broad in 1920 and 1921 respectively. Podds also owned the early motor drifter *J.A.P.*, which was built in 1931 and named for Jane Anne Podd. She fished on until 1967. Diesel Trawlers' motor smacks included the *Dorando Pietri*, built at Oulton Broad in 1908, the *Boy Clifford* LT 1202, built at Lowestoft in 1920, and the *Helping Hand* LT 1239. These were among the more powerful motor smacks, having 135 hp engines. The *Dorando Pietri* was topically named after the 1908 Italian Olympic runner. Inshore Trawlers owned the *Dusky Queen* LT 895, built at Oulton Broad in 1920, and the *Golden Lily* LT 1186, built at Lowestoft in 1912.

The *Purple Heather* LT 249 was built in 1921 by John Chambers at Oulton Broad to replace a smack of the same name that was sunk by a German submarine in 1915. In 1933 Richards Iron Works fitted her with a 100 hp Crossley diesel and converted her to a full motor smack. *Rosary* LT 73 was built in 1924 by Rye boatbuilders C. & T. Smith. In 1934 she was fitted with a 125 hp Crossley engine. After being sold to Aberdeen, she fished until 1952.

Several of the motor smacks had very long careers. The *Flag Jack* worked until 1952, when she was lost by a motor explosion, and the *Master Hand* was still working from Plymouth in the 1950s, registered in Brixham as BM 43.

As well as the English conversions to motor trawlers, many smacks were converted to power by foreign owners, including the Belgian *Pierre* O 281 (BM 1 *Superb*), the *Charles Yvonne* O 46 (*Sunbeam* LT 19), the *Jean Andre* B 30 (*Pentire* LT 742) and the *Ocean's Gift* O 152 (*Ocean's Gift* BM 223).

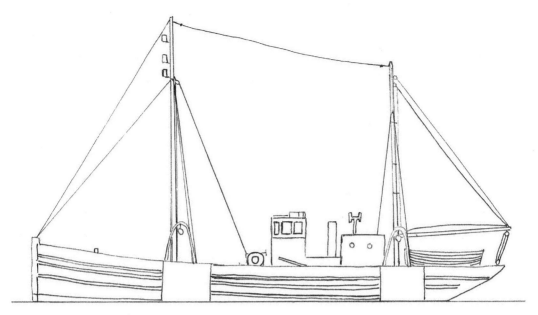

Lowestoft motor smack.

Built at Rye, Sussex, in 1924, the 70-foot smack *Holkar* LT 18 was fully converted to a motor trawler with an Allen engine. She fished from Lowestoft until 1962 and, after being briefly used as a cable layer by Post Office Telephones, was sold to Swansea.

British Motor Fishing Vessels

LOWESTOFT AND YARMOUTH MOTOR DRIFTER

The first English steam drifter, the *Consolation* LT 718, was built in 1897. The last to be built in the UK was the *Wilson Line* KY 322 by A. Hall & Co. of Aberdeen in 1932. The Dutch built economical motor drifters (motor loggers) from the First World War and the Belgians successfully converted several British-built steam drifters to motor trawlers during the 1930s, while the Scots developed very economical wooden motor drifters. Some early English motor drifters were not a success, including the *Thankful* LT 1035 of 1907 and the *Veracity* LT 311 of 1926. Successful pioneers were the *J.A.P.* LT 245, built with two 70 hp Deutz diesels in 1931, and the *Tojo* LT 69, which dated from 1905 and was converted from steam to motor with a 200 hp Mirrlees diesel in 1935.

By 1945 it was clear that the steam drifters were no longer economical to run and many of them were worn out after their arduous service in the war. Early motor drifters built after the Second World War were the wooden *Equity I,* better known as the *Madame Prunier* LT 343, from Brooke Marine, and the steel *Dauntless Star* LT 371, from Cochrane of Selby. The *Dauntless Star* promptly won the 1948 Prunier Herring Trophy for a top shot of 267 crans (a cran was about 28 stone or four baskets), which sold for £905 on 9 November.

But the pattern for the last generation of Lowestoft and Yarmouth motor drifters was set by the *Frederick Spashett* LT 128, which was built for the Small Group by Richards Ironworks of Lowestoft in 1949. She was followed over the next decade by a whole series of broadly similar steel drifter/trawlers, including the *Henrietta Spashett* LT 82, *George Spashett* LT 184, *Harold Cartwright* LT 231, *Ocean Sunlight* YH 167, *Ocean Starlight* YH 61, *Young Elizabeth* LT 375, *Young Duke* LT 387, *Dick Whittington* LT 61, *Norfolk Yeoman* LT 137, *Ocean Dawn* YH 77, *Ocean Surf* YH 107, *Ocean Crest* YH 207, *Ocean Trust* YH 327, *Ethel Mary* LT 337, *Valiant Star* LT 277 and *Suffolk Warrior* LT 671. These craft were seen at every herring fishery between Shetland and Lowestoft and some of them came west to Newlyn for the spring mackerel voyage. Many of the more recent steam drifters were converted to motor, often on similar lines to the new vessels, with streamlined wheelhouses and funnels. Among them were the *Fellowship* LT 246, *Eager* LT 1166, *Wilson Line* KY 322, *Kindred Star* LT 177, *Sedulous* LT 56 and *Thrifty* LT 152. Several of these were mainly engaged in trawling.

The *Frederick Spashett*, powered by a 231 hp Ruston engine, was 81 feet long with a 21-foot beam and a depth of 9 feet. She was one of seven drifters sold to South African owners in 1965. The *Dick Whittington* won the 1961 Prunier Trophy on 23 October with 274 crans under Skipper Leo Borrett. Other big shots were landed by the Scots drifters *Xmas Star* FR 87, with 196 crans, and *Wilson Line* KY 322, with 170 crans.

The *Dick Whittington* LT 61 was one of the drifters to fish the Irish herring season from Dunmore East. She was sold to Italy in 1968. Skipper Stanley Hewett in Bloomfield's *Ocean Starlight* YH 61 won the 1961 Trophy with a shot of 294 crans. One of the drifters bought by Smalls of Lowestoft, she was sold to Holland in 1967 but subsequently returned to the UK to work as an oil rig stand-by vessel, as did several of her sister drifters.

The Ocean-named drifters were owned by the famous Bloomfields company of Yarmouth that was well known for the soap names of some of its steam drifters (*Ocean Vim* YH 88, *Ocean Lux* YH 84 and *Ocean Lifebuoy* YH 29), which were derived from the firm's connection with Unilever. In 1963 the remaining Yarmouth Ocean drifters were sold to Small of Lowestoft. The age of the drifter was almost over as a result of overfishing and the introduction of pair trawling for herring. 1968 was the last year in which a drifter fished from East Anglia.

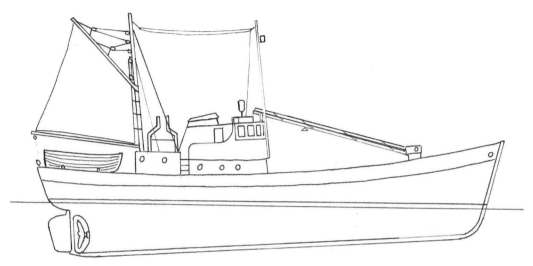

Lowestoft and Yarmouth motor drifter.

Young Elizabeth LT 375. (Watercolour courtesy of Michael Pellowe)

British Motor Fishing Vessels

97-FOOT MFV DRIFTER

Of the four classes of MFVs built as Naval auxiliaries during the Second World War, the biggest were the 97-footers. They had a 22-foot 3-inch beam, an 11-foot draught aft and displaced 200 tonnes. Powered by Crossley 240 hp diesels, they were reputed to be ungainly beasts with poor manoeuvrability, but were popular with sailors as they could ferry ashore 400 liberty men in one trip (about half the crew of a cruiser). They were massively constructed, which accounts for their longevity. A total of 72 97-foot MFVs were built, compared with 208 75-footers, 342 65-footers and 227 of the smallest 50-foot class.

After the war, many were sold for fishing, with large groups going to Lowestoft and Brixham. Most were adapted for trawling but some were drifter/trawlers, or just drifters. MFV 1506 was built by Richards of Lowestoft in 1944. In 1948 she was sold to Claridges of Lowestoft and was converted to a drifter/trawler. She was named *Eta* LT 400 after the Ala-class steel trawler that was mined during the war. In 1956 she was sold to Putfords and was renamed *Arduous*. Under Putfords' ownership, she was used exclusively for trawling, and in 1958 she was re-engined with a 360 hp Ruston engine. The *Arduous* finished her working career as an oil rig stand-by vessel before being sold for leisure use in 1969.

Built at Totnes on the River Dart in 1946, MFV 1536 became the Grimsby trawler *Harrowby* GY 531. Sold to Star Drift Fishing of Lowestoft, she became the drifter *Vesper Star* LT 317. She was sold in 1961. Star Drift's *Friendly Star* LT 318 was built at Plymouth by G. Hall in 1946 for the Admiralty as MFV 1554. In 1947 she was sold to Shire Trawlers and became *Firsby* GY 494 before her purchase by Star Drift Fishing as a drifter in 1951. She had a ten-year career as a Lowestoft drifter until she was sold to Agadir in Morocco in 1962. A similar story was with the *Starlit Waters* LT 97, which was ordered by the Admiralty as MFV 1577 and completed at Lowestoft in 1948. Bought by F. E. Catchpole, she was subsequently part of Star Drift's drifter fleet before working as a trawler for Lowestoft Fishselling Co. She was sold from Lowestoft in 1965.

One of the best known 97-foot MFVs, the *Betty Leslie* LK 497, had a long and varied career. She was built by Richards of Lowestoft in 1943 as MFV 1504 with the usual Crossley 240 hp diesel. In 1948 she was registered in Lerwick as the *Betty Leslie*. The biggest drifter in Shetland, she made the annual autumn voyage to Lowestoft for Home Fishing under Skipper George Leslie. She also worked long lines as far away as Iceland. In 1948 the Herring Industry Board (HIB) arranged for a group of British fishermen to study purse seining in Norway. In 1950 the HIB hired the *Betty Leslie* and converted her for purse seining with a moveable platform aft where the net was worked. However, the purse seine trials seem to have been inconclusive. In 1960 she was sold to the Boston Deep Sea Fishing Co., was renamed *Boston Mosquito* LT 373 and re-engined, and was completely refitted as a trawler. She ended her working life as a safety standby vessel.

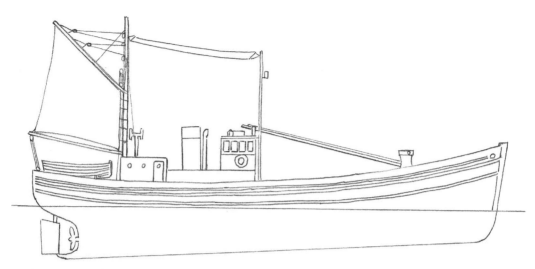

97-foot MFV drifter.

The 97-foot wooden MFV drifter *Quiet Waters* LT 279 was built at Par, Cornwall, and was completed in 1946. Owned first by F. E. Catchpole and then Gilbert & Co. of Lowestoft, in 1961 she was one of the Lowestoft drifters that sailed to Newlyn to fish the Westward Voyage for mackerel. The local newspaper commented on the mackerel season: 'East Coast mackerel drifters are not doing too badly at all with landings fairly regular and prices holding steady.'

British Motor Fishing Vessels

97-FOOT MFV TRAWLER

Many trawlers were damaged or lost during the Second World War and several owners snapped up the 97-foot MFVs as they became available. Well known on the West Coast was the *Heather George* SA 29 of Swansea. She was distinctively painted pale blue with yellow masts and funnel. Built at Lowestoft by Richards Ironworks in 1945, she frequently worked the Smalls and Trevose grounds, where she was often surrounded by large numbers of Ostend and Zeebrugge trawlers

Torbay Trawlers of Brixham acquired a large fleet of these vessels, including the *Agnes Allen* BM 21, *David Allen* BM 10, *Elijah Perrett* BM 15, *Iago* BM 20, *Moonlit Waters* BM 23 and *William Allen* BM 137. Gloucester Trawlers Ltd owned the *Roger Bushell* BM 76. In April 1950, the *Roger Bushell*'s trawl came fast while she was working the Wolf Grounds. One of her warps parted, resulting in three crew being injured, who had to be landed at Newlyn. MFV 1557 was built by East Anglian Constructors at Oulton Broad in 1944. In 1948 she was bought by Max Aitken and named *William Rhodes Moorhouse* LO 496 after a Great War airman who was awarded the Victoria Cross.

In 1958–59, several Lowestoft 97-foot ex-MFVs were radically refitted with modern bridges and raised forecastles, and were equipped with more powerful engines. The Colne Group acquired the *Red Snapper* in 1952, which was later renamed the *Snapper* LT 303. After modernisation she served until 1962, when she was sold to South Africa. Similar refits were carried out on the *Yellowfin* LT 282 (ex-*Alorburn*), *Yellowtail* LT 326 (ex-*Maravanne*), *Boston Mosquito* LT 373 and *Kirkley* LT 225 (drawing).

Stevensons of Newlyn made a success of their 75-foot MFV trawlers and in 1957 bought the 97-foot former *Sybella* MFV 1564 and converted her to the trawler *St Clair* PZ 199. She quickly established a reputation of staying at sea when bad weather drove the smaller vessels into port. In March 1961 she landed a 1,400-stone catch, which earned a record £1,000 on the Newlyn market. Following the success of the *St Clair* PZ 199, Stevensons of Newlyn acquired three more of these vessels from Torbay trawlers in 1962: the *Agnes Allen* BM 21, renamed *Elizabeth Ann Webster* PZ 291, the *David Allen* BM 10, renamed *Elizabeth Caroline* PZ 293, and the *Elijah Perrett* BM 15, which became the *Marie Claire* PZ 295.

These three vessels had the most remarkable, and certainly the longest careers of all the 97-foot MFVs. They fished successfully as traditional side trawlers during the 1960s until, following the lead of the Brixham beamers *Sara Lena* and *Scaldis*, it was decided to convert the *Elizabeth Ann Webster* for beaming at Dartmouth ship builders Philip & Sons. She was decked over aft and fitted with a gantry and derricks in front of her new modern wheelhouse. The *Elizabeth Ann Webster* emerged from her refit looking a completely new ship, the first of many beamers to be owned in Newlyn. Next was the *Elizabeth Caroline* built by Richards in 1945 as MFV 1568. She too was fitted with a gantry for her beam derricks, a new raised forecastle, wheelhouse, trawl winch and 495 hp main engine. After being laid up at Newlyn, in 1986 the *Marie Claire* was refitted by local boatbuilder Raymond Peake and the Maaskant shipyard in Holland. She too received a new wheelhouse, gantry and 600 hp Dutch engine, and was set for a new career in the twenty-first century. The story of these three vessels is a testament to their original builders, who built well, but also to the imagination and enterprise of their owners.

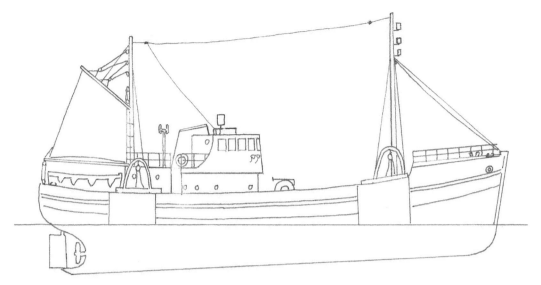

97-foot MFV trawler.

Beamer *Elizabeth Ann Webster* PZ 291 on the slip at Newlyn in 1974. (Photograph Tony Pawlyn)

British Motor Fishing Vessels

EARLY DANISH SEINER

The Danish seine was invented in Limfjord in 1848 by Jens Væver, who quite rightly became a fishing hero. Beach seines were, and still are, used in many countries. The net is shot out from a beach in a U shape by a small boat and is then hauled back to the beach by long ropes, hopefully bringing some fish with it. Jens Væver had the idea of taking the net to sea and shooting it from an anchored boat. Originally, the fishing boat anchored, and the net was set from a small rowing boat. The fishing boats were fine ketch-rigged kutters, fitted with wet wells (viviers) for keeping the catch alive. Danish fishermen were quick to adopt motors, first of all for hauling in the seine net ropes. Some of them found another use for this auxiliary engine, however; a drive shaft over the stern drove a propeller by a bicycle-type chain, and this arrangement could give the boat a speed of 2 to 4 knots in calm weather. Next, the kutters' boats, which shot their seines, were fitted with engines. These big double-ended boats were towed, or were sometimes taken aboard. Motors worked so well that the small boats were soon redundant and a class of small powered motor boats, known as sharks, largely replaced the big seine kutters.

Like the trawl, the seine is a stocking-shaped net, but unlike the trawler the anchor seiner does not tow her net along. The boat drops anchor with two buoys on it. From the buoys she shoots one length of warp, then the net, then the second warp back to the buoy. The set is made in a roughly triangular shape, with the net in the middle of the base of the triangle. When the seiner returns to her anchored buoys, both warps were hauled in by the winch, which was fitted with a coiler to coil down the ropes. The crew stacked the coils on the deck, ready for the next shot. As the ropes were hauled, they swept over the seabed, causing a cloud of sand (so they only fished in daylight), which drove the fish in front of the net. As the winch and coiler did the work, Danish seining was very economical on fuel, and could thus be done by quite small skipper-owned boats.

Because the seiner anchored in the middle of the North Sea, often in poor weather, she had special anchor gear made up of a heavy anchor and chain, a length of rope (synthetic in recent times) and a long wire interrupted by another length of chain, supported by two huge round buoys. Her winch had a separate drum for working her anchor wire.

The Danish seiners from Frederikshavn, Skagen, Grenaa and Esbjerg began fishing in the North Sea after the First World War and often landed their catches at Grimsby. They were so successful that they were soon copied by British fishermen, who were grateful for an alternative fishery at a time when the herring industry was in deep depression with the failure of its markets in Germany, Poland and Russia.

A typical 50-foot seiner from this time could have a slow-running 48 hp Alpha engine and four to six crew. The wet well was abolished and the fishroom was fitted with shelves where the catch was iced, enabling them to make fortnight-long trips. The winch was driven by a belt and drive shaft from the fore end of the engine, and was fitted with a two-speed coiler. Danish seiners were strongly built of oak, with full lines forward and a high sheer to keep them dry while at anchor. The engine was aft, the cabin forward and the winch and fishroom were amidships.

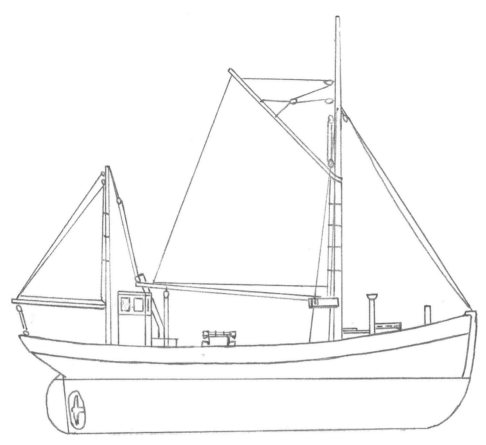

Early Danish seiner.

Danish seiner.

British Motor Fishing Vessels

STEAM SEINERS

Perhaps one of the quirkier aspects of seining in the UK is the story of the steam seiners. With the huge success of the Danes landing their catches at Grimsby, in 1920 six Scots drifter skippers bought seine nets and winches at Grimsby from the Great Grimsby Coal Salt & Tanning Company. Seine net winches came from the local Chapman company and coilers were imported from Denmark. Once ready, these drifters started seining for haddock, cod, plaice and lemon soles in the Moray Firth. Boats from Buckie, Lossiemouth and other Moray Firth ports also took up seining.

As with all new ventures, there was much to be learned. Some of the earliest steam seiners improvised by hauling their ropes with their steam capstans, which were fitted with chocks to separate the ropes. To begin with, the seine ropes were coiled by hand, in the same way as the footlines of drift nets. They were coiled in the same place, the rope locker, which was forward, below decks. However, it was soon realised that proper winches and coilers were needed, and the coils were then laid out on deck. Seine net catches rose from 760 tonnes in 1921 to 2,500 tonnes the following year and 4,360 tonnes in 1924.

In 1922 Sam Franklin formed the Grimsby Seine Fishing Co. Large numbers of brand-new steam drifters known as standard drifters, which had been built as auxiliaries for the Royal Navy, were available. Powered by 42 hp triple expansion steam engines giving a speed of just under 9 knots, these were of 96 gross tonnes and were 94 feet long, with a 20-foot beam and a draught of 10 feet. The standard drifters were all named after meteorological phenomena. Franklin's *Shade* GY 186, *Blacknight* GY 188, *Cloudarch* GY 187 and *Cat's Eye* H 316 (originally *Sirocco*) were built by Colby Brothers at Oulton Broad, while their *Swirl* GY 189 was built by Chambers of Lowestoft, the *Maelstrom* GY 15 and *Greynight* GY 141 were built at Lymington and the *Whitenight* GY 88 was built at Barton-on-Humber.

Franklins eventually acquired a fleet of eighteen vessels, which fished up to the Second World War. Because of the hardship caused by the lack of markets in the herring fishery, Franklins were able to take on herring fishermen from Lowestoft, Yarmouth and the East Coast of Scotland ports. Fish were plentiful after the fishing grounds had restocked during the First World War, when the steam trawler fleets had been requisitioned for war service. In the 1930s the fishing industry was affected by the worldwide Depression, and Franklins alone persevered with their steam seiners until the Second World War. Such was the initial success of Franklins that they were soon followed by other local owners – Thomas Bascombe and Garratt & Jeffs.

The Grimsby steam seiners successfully fished mainly for haddock. They usually worked trips of about a week in length with a crew of eight.

Other steam drifter seiners were the *Jacob George* YH 176, *Nelson* LT 515 and *Girl Norah* LT 1137. The *Girl Norah* was lost on the French coast near Calais on 2 December 1936. Her nine crew put off in her small boat, but tragically eight of them were lost – the single survivor being rescued by a small French fishing boat.

There were different kinds of seines for haddock or plaice. They were tarred and fitted with corks and hollow glass floats at the top, and leads at the bottom. Although motor boats were to prove more suitable as seiners, the steam seiners were an important part of the story of seining both in England and Scotland.

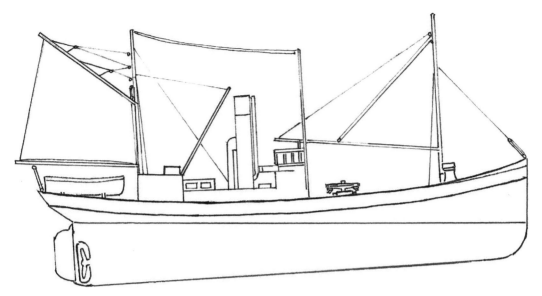

Steam seiner.

Steam seiner *Jacob George* YH 176. Her winch and coiler are in front of the wheelhouse. The *Jacob George* was built in 1910 by the J. Chambers shipyard at Lowestoft. While herring drifting, she was wrecked at Eyemouth, Scotland, and her crew were rescued by breeches buoy. She survived this incident and the Second World War and was finally scrapped in 1949.

British Motor Fishing Vessels

DANISH SEINER

In the 1930s, limits were put on foreign landings at Grimsby, but some fortunate Danish skippers were licensed to work from the port and fished there for whole seasons. Local owners began to acquire their own motor seiners, such as the *Creag Mhor* GY 17, *Clavis* GY 347 and *Geneara* GY 382. Local owners Marstrand bought the new *Gipsy Love* GY 204 from Sweden and had the *Girl Pat* GY 176 built at Oulton Broad. Powered by a 120 hp engine, the *Girl Pat* was 66 feet long with an 18.7-foot beam and a depth of 8.7 feet. Unfortunately for her owners, she achieved fame by being hijacked by her skipper and sailed to the Caribbean in 1936.

In 1940, Denmark was invaded by the Germans, and many Danish fishermen escaped to the UK with their vessels. While many boats were requisitioned by the Royal Navy, others fished from Fleetwood on the West Coast. In 1946 there were only a dozen seiners at Grimsby, but such was their success that the fleet had grown to 140 (including pair trawlers) by the late 1970s. Many of their owners and skippers were Danish fishermen who settled in Grimsby and Hull with their families and continued to build their boats in Denmark.

During the post-war years several Grimsby seiners left their North Sea grounds over the winter months and fished in the Moray Firth. Their skippers and crews got to know the highly reputed local Scots boatyards and used them for repair work. In the 1950s and '60s many anchor seiners were built for Grimsby and Hull by the Buckie yards of Thompsons, Herd & Mackenzie and Jones, aided by the White Fish Authority's Grant & Loan scheme. With the contraction and eventual disappearance of the deep sea trawler fleet as Iceland and other countries extended their territorial waters, several trawler owners diversified with seiners. The St Andrews Fishing Company built the *Viborg* H 33 and the *Nordborg* H 35 in 1957 at Herd & Mackenzie. In 1974 Boston Deep Sea Fishing of Hull built the *Arnborg* H 272 and the *Vikingborg* H 285 at the same yard.

A typical seiner from the 1970s was 60 feet long with 18-foot beam and a depth of 8.5 feet and was powered by a 150 hp diesel. The galley and accommodation for three to four crew was forward, while the skipper usually had his cabin aft, in the wheelhouse. The coiler and the manhandling of coils of rope were replaced by two large rope reels forward on the port side, which wound on the warps as they were hauled by the winch. The introduction of rope reels was a big contribution to fishing efficiency and crew safety. There was often an extra reel for turning the warps around so that they wore evenly. The first anchor seiner to be built equipped with rope storage reels was the *Arnborg*, which was launched in 1974. Instead of the seine net being hauled aboard manually, a hydraulic power block was fitted on the stern.

While the anchor seiners were being modernised, similar vessels, instead used for gill netting, were fitted with raised foredecks and shelter decks. Space needed in the bows for handling the anchor made it difficult to fit Danish seiners with foredecks. However, when they *were* fitted, a section could be raised to give the crew access forward.

In 1968 the Grimsby seiners began pair trawling for white fish – a technique that brought immense success. With the decline of the deep sea trawler fleet, the seiners and pair boats made a vital contribution to landings at Hull and Grimsby.

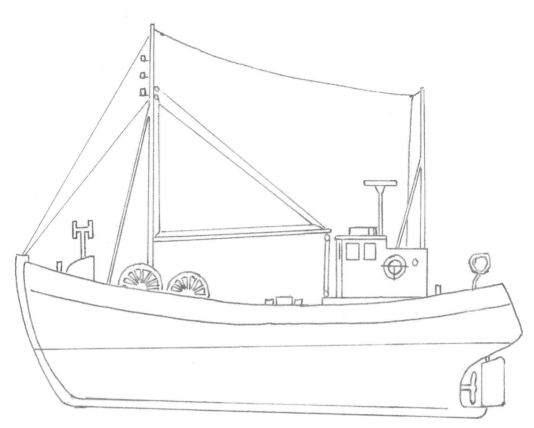

Danish seiner.

The 58-foot Grimsby seiner *Foursome* GY 536 was built at Faaborg, Denmark, in 1943. Powered with a 134 hp engine, her winch is amidships, while her rope reels and anchor buoys are forward.

British Motor Fishing Vessels

COBLE

Perhaps the best known British traditional boat, the coble, belongs to the coasts of Northumberland, Durham and Yorkshire. Of a unique shape, it was built around a centre plank called the ram, which was flat at the stern and came down to a deep bow. There was no visible keel aft; instead, there were two side keels called drafts for grounding on the sand. Designed as a beach boat, the coble's high bow kept her dry and her deep forefoot gripped the waves when she was being launched, while the two drafts held her steady when she was beached stern-first at the end of her day's work. Clinker-built on sawn oak frames, her top plank sloped inwards towards the stern. The sailing coble was rigged with a dipping lugsail, a jib and bowsprit. Since she drew very little water aft, her deep rudder acted as a centre board. Working cobles are getting rare, but Bridlington prides itself on its little heritage fleet of sailing cobles, some which have been built recently.

When motors were fitted in cobles, there was a problem: where to put the propeller? Boatbuilder William Clarkson of Whitby devised a tunnel stern for his coble *Topsmoz*, which was built for Skipper Halenshaw of Filey in 1935. She was 27 feet 7 inches long with a very modest 7-foot 4-inch beam and a maximum draught of 1 foot 10 inches. She was planked in larch on oak frames spaced at 1 foot 4 inches between centres. Her engine was a petrol-fuelled four-cylinder Thorneycroft, which drove a two-bladed propeller of 18 inches in diameter.

The smart white-painted *Silver Coquet* BH 63 was built by J. & J. Harrison of Amble for the Henderson family of the same port. She was powered by a Petter diesel and was fitted with a line and pot hauler. Amble's Northumberland cobles worked salmon nets and crab pots. When line fishing, they shot up to six lines, each of which was fitted with about 700 hooks. Her foredeck was covered by a canvas dodger.

The Redcar coble *Daisy Ellen* MH 208 built in 1968 by Gordon Clarkson of Whitby for Jimmy and Mark Thompson was a much heftier 34-foot-long craft which worked lobster pots and gill nets for salmon and sea trout. Powered by a Parsons Porbeagle 60 hp diesel, she was fitted with a radio and echo sounder, a canvas dodger over her foredeck and a tripod foremast to hold a derrick, deck lights and aerial. Like other inshore craft, cobles have been progressively modified to suit local fisheries. Many are fitted with hydraulic net haulers, others with vee wheel (slave) pot haulers, and some of the biggest have trawl winches and gantries. Many of the larger cobles have forward wheelhouses, often with a radar on top.

Fishing cobles continued to be built almost until the end of the twentieth century, and some are still at work, but in declining numbers. They are fast becoming heritage boats rather than working ones, but their iconic status as valuable representatives of local maritime history is widely recognised.

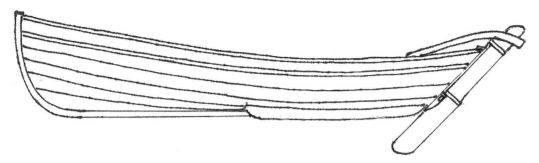

Coble.

Staithes cobles *Coronation Queen* WY 75, *Sea Lover* WY 99 and *Golden Crown* WY 78 in the early 1950s. As might be guessed from their names, the *Coronation Queen* and *Golden Crown* were built in the year 1953. The *Coronation Queen* was built by Harrisons of Amble. Sold to Whitby in the 1970s, she fished until 1995. Built by Harrisons for George Hanson the following year, the *Sea Lover* was powered by a 15 hp Kelvin petrol paraffin engine. The *Golden Crown* was built by William Clarkson of Whitby for Richard, Matthew & Francis Verril to replace the *Star of Hope* WY 174, which had been wrecked. She was 30.5 feet long with a beam of 8.4 feet and a depth of 3.3 feet. Powered by a Petter 10 hp diesel, she hauled her gear with a Hyland hydraulic capstan aft. These cobles worked about 150 pots while shell fishing for much of the year, and went long-lining for cod during the winter.

British Motor Fishing Vessels

YORKSHIRE DOUBLE-ENDER

In contrast to the English square-sterned coble, with its sawn timbers and wide clinker planks, the twentieth-century double-ender typically has narrow planks and steamed timbers, enabling her boatbuilder to develop the design of a hull intended to be beached bow-first. But there have been several different kinds of double-enders in the coble family.

The Scarborough and Bridlington pointed-sterned coble of the start of the twentieth century was very much a coble with sawn frames, wide planks and a tall dipping lug sail on a sloping mast, and was very similar to her better-known sister. These cobles had a fine run aft and were thought to be safer boats when running before a following sea.

The big nineteenth-century herring cobles, known as mules, were hefty craft with pointed sterns and fuller lines. A typical Filey herring coble was 42 feet long with a 13-foot 6-inch beam. Like the smaller craft, she was rigged with one mast, a jib and dipping lugsail with several rows of reef points. Her long rudder reached well below the level of her keel.

Seahouses in Northumberland had a big fleet of cobles and a group of distinctive decked mules with low wheelhouses amidships in the 1930s and '40s, including the *Providence* BK 142, *Respect* BK 162, *John Wesley* BK 14, *Blossom* BK 40 and *Nelsons* BK 445. These boats went herring drifting and seining. These mules originated in Scotland and were like miniature fifies.

In Yorkshire, some landings like Filey and Redcar had a mixture of cobles and double-enders, while at Runswick, double-enders were more popular. While the coble has to be turned stern-on to the shore and her rudder removed before she is beached stern-first, the double-ender is steered straight in bow-first. Like the coble, she has her tractor and trailer waiting to take her up the beach. Many skippers preferred double-enders as they were cheaper to build and carried more gear than a coble.

A 32-foot double-ender built by William Clarkson of Whitby for local skipper J. Harrison in 1937 followed an almost identical *Silver Line*, which was built the previous year. She was fitted with a capstan aft for handling her crab pots and long lines and was also licensed to carry twelve passengers. She had fine ends and a narrow beam of only 8 feet 6 inches – slightly more than a quarter of her length – but had more sheer than modern double-enders. Her keel was of Columbian pine and her stem, stern, keelson, gunwales and timbers, spaced at 16-inch centres, were of English oak. She was planked with larch and powered by a 15 hp Kelvin Ricardo petrol/paraffin engine, which drove a 21-inch two-bladed propeller. Her capstan was driven off the fore end of the engine, her fuel tank was in the after locker, and she carried a mizzen mast and steadying sail. Although there was provision for a foremast, it was thought unlikely that it would be used. The *Silver Line* was almost identical but for her engine – a 20 to 25 hp Thornycroft petrol/paraffin engine driving a 16-inch three-bladed propeller to give a speed of 8 knots.

The *Triumph* SH 51, built thirty years later by Gordon Clarkson, had a much higher beam ratio, a fuller bow and stern and less sheer. Built for Filey Skipper Stan Cammish, who came from an old established local fishing family, she was 24 feet long with a 7-foot beam and was powered by a Petter 16.4 hp twin-cylinder water-cooled engine. Smartly maintained with varnished topsides and a white bottom, she worked long lines between Filey Brigg and Flamborough Head for spragg (cod), codling and haddock. The White Fish Authority's Grant & Loan scheme (which helped fishermen build boats), as well as the extending of British territorial waters in 1964, were a shot in the arm for fishermen, and many fine boats were built as a result.

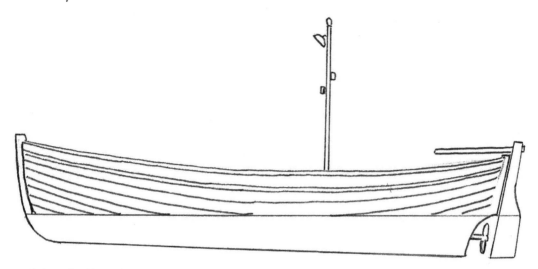

Yorkshire double-ender.

Double-enders at Redcar, including the *Mizpah* and *Pretoria*.

British Motor Fishing Vessels

LINERS AND KEELBOATS

Lining on the Yorkshire coast was a fiddly and labour-intensive business, the lines being cleared and baited ashore with mussels, whelks or limpets. But liners could fish rough grounds where trawlers could not venture. While the Bridlington fishermen were well-known liners, the keelboats (i.e. decked boats) of other ports, like Whitby and Scarborough, diversified into seasonal potting, herring drifting and seining. From the 1960s, trawling became more important.

In 1932 the Alexander Aitken yard at Anstruther completed the liner *Olive Branch* GY 409 for Sam Chapman & Sons of Grimsby. 50 feet long with a 15-foot beam and a draught of 5 feet 6 inches, the *Olive Branch* was driven by a 60/70 hp Kelvin Ricardo engine that started on petrol and ran on paraffin. Her engine room was aft, the fish hold amidships, and the cabin, with seven bunks and a coal cooking stove, was forward. Her hold was fitted out with an ice locker, fourteen pounds and shelves. Since she was intended for lining, the small hatch over her hold gave her clear decks. There were no drift net hatches. She was steered by a worm gear (a screw-threaded bar from the back of the wheel to the rudder head).

The 49-foot *Elsie Mabel* GY 330, built by Humphrey & Smith of Grimsby to a design by A. G. Dalgarno in 1937, was of the same layout as the *Olive Branch,* but with an elegant canoe stern. Claimed to be the first liner built in Grimsby for forty years, she had higher bilges, a deeper draught and more flare than Scots-built keelboats. At a time when diesel power was becoming standard, she was fitted with a 55 hp Thornycroft six-cylinder paraffin engine with a 4:1 reduction gear.

The cruiser-sterned *Our Confidence* H 461 was built in 1949 by Clapson & Sons of Barton-on-Humber for Skipper John Newby of Bridlington. She was powered by a 96 hp Dorman diesel and was also fitted with a ring net winch. During the winter she lined for cod and in the summer for dogfish – a firm favourite with fish and chip shops as 'rock salmon'. The lines were shot at night. Each line was coiled and baited on to a basket work skep. The *Our Confidence* worked twelve lines, each of which was 315 fathoms long. All twelve of the lines were joined together to be shot as one line. The hooks were spaced 8 feet 6 inches apart. The lines were hauled by hand and the fish gaffed aboard. Not having been squashed together in a trawl, line fish were of high quality and fetched good prices, but the expenses of lining were high. The *Our Confidence* had five crew but an equal number were kept busy in her baiting shed ashore, clearing and baiting her lines.

Like Whitby and Scarborough, Bridlington later became a trawling port, but today the town has one of the biggest crabbing fleets in the UK.

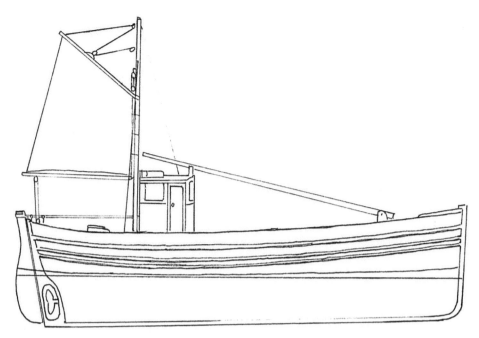

Liners and keelboats.

The Whitby keelboat *Galilee* WY 68 was built by W. & G. Stephen of Macduff in 1932 as one of dozens of Whitby keelboats to hail from Scots boatyards. 44 feet long with a 14-foot beam, the *Galilee* was driven by a 44 hp Kelvin K2 diesel. During the 1940s and '50s, many Scots ring netters fished herring from the Yorkshire ports. Several local keel boats tried ring netting, including the *Galilee*, *Endeavour* WY 1, *Provider* WY 71, *Success* KY 211 and *Wayside Flower* H 65.

British Motor Fishing Vessels

FROM SAIL TO POWER IN SCOTLAND

A Peterhead lugger running for harbour; they must have had nerves of steel! When fishing, she lowered the foremast into the crutch beside the mizzen mast and hauled her nets over the stern with the help of her capstan aft.

Cullen Harbour packed with zulus and steam drifters *c.* 1910–13. Before the First World War, the Scots East Coast ports rapidly replaced many zulu and fifie luggers with steam drifters. Left to right are: steam drifter *Guiding Star* BF 304, zulus *Jeremiah* BF 7 and *Smiling* BF 772 (with a broken mizzen mast), and steam drifters *Herald* BF 908 and *Sunnyside* BF 167.

The Cullen steam drifter *Crannoch* BF 877, seen in 1934, was one of hundreds built in the early twentieth century for ambitious Scots fishermen. Originally the *Jeannie Gilchrist*, she was built of steel at Montrose in 1907 for Findochty owners. Requisitioned for Naval service from 1917 to 1919, she was sold to Cullen and was renamed in 1920. The *Crannoch* fished until 1938.

The St Monans motor fifie *Elspeth Smith* KY 38, which was built locally in 1904. Her mizzen mast is in front of the wheelhouse. Her capstan is now forward, while amidships is a roller used for handling her nets. On the left, newly painted net buffs are hanging up to dry.

British Motor Fishing Vessels

MOTOR ZULU

The traditional fishing boats of the East Coast of Scotland were the double-ended scaffie, with a sharply raked bow and stern, and the fifie, where the bow and stern were almost vertical.

In 1879 boatbuilder William Campbell of Lossiemouth built the lugger *Nonsuch* with the straight bow of a fifie and the sloping stern of the scaffie. This new kind of boat was called a zulu, after the war in South Africa. The zulu was a success, regarded as faster and more manoeuvrable than the fifie and safer and more stable than the scaffie. When the steam capstan arrived in the 1890s, zulus were built up to 80 feet long, as the capstan was used to haul in the nets over the stern by their bush rope, raise the massive foremast, hoist the huge sails and land the catch.

These magnificent, speedy luggers had barely twenty years of popularity before being superseded by the steam drifter in the early twentieth century. The progressive Scottish fishermen were also quick to adopt motors. The zulu's raked stern was not suitable for fitting a propeller, so a wide stern post was often built on to make room for the propeller aperture, which can be seen in the drawing. A few motor zulus were built, including the *Spindrift* WK 177, *Surprise* LK 477 and *Evangeline* FR 163. One of the first big zulus to be fitted with a motor was the 80-foot *Mother's Joy* BF 892 of Gardenstown, which was fitted with a Fairbanks Remington paraffin engine in March 1909. In the autumn she sailed for the East Anglian herring season and had a successful voyage working from Lowestoft. In 1911 the Rosehearty zulu *Heather Bell* was fitted with two Kelvins by Forbes of Sandhaven and began a long career as a motor boat. The 79.6-foot *Avoca* BCK 293 was built at Buckie in 1904 and fitted with a Gardner 55/60 hp engine in 1918. She fished until 1952.

The last big zulu to work was the *Research* LK 62, which was built at Banff in 1901. She was fitted with a small auxiliary engine around the start of the First World War, in 1914. In 1922 this was replaced by two 50/60 hp Kelvin petrol/paraffin engines. She was bought by Skipper Robert Polson of Whalsay and partners in 1935 and fished from Lerwick, mainly at the herring, until 1968. She was a top herring drifter. In 1961 she landed her biggest shot of 210½ crans. After many problems, she was eventually conserved by the Scottish Fisheries Museum at Anstruther, Fife.

The small zulu *Violet* FR 451 was built by James Noble for Alexander Stephen of Fraserburgh in 1911. She fished with lines and herring drift nets. She was soon fitted with a 15 hp Kelvin, which was afterwards replaced by a 30 hp engine and then with a 48 hp Gardner in 1936. Beautifully maintained, she was finally sold out of fishing in 1975 and became a yacht in the USA. The Stephen family also owned the similar *Vesper* FR 453, which was also built in 1911, and which was sold from Fraserburgh to Pittenweem in 1970, only ceasing fishing in 1988. The zulu *St Vincent*, built in 1910 by Stephen of Banff, was completely restored to her original lugger rig at Arbroath in 2011 after a long career fishing from Eriskay, Wick and Shetland.

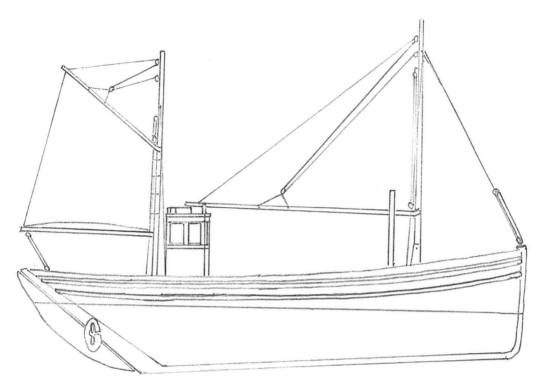

Motor zulu.

The zulu *Columcille* OB 147 was built at Arisaig in 1912 as the *Virgin*. She was 45 feet long with 15-foot beam. She ended her career as a creeler working from Mull. In 1964, she sank north-west of Jura.

British Motor Fishing Vessels

MOTOR FIFIE

With an almost vertical stern post, the fifie was more readily converted to motor power. Usually, a hole was cut in the rudder to make room for the propeller. The first big fifie to be motorised was Skipper H. Swanston's 65-foot *Maggie Jane's* BK 146 of Eyemouth, which was fitted with a 55 hp Gardner in 1907. Unusually, her steam capstan was replaced by a Gardner motor one. (Motor boats continued to be fitted with steam capstans into the 1920s.) After three years of successful use, her engine was replaced by another Gardner in 1910. Sailing to the East Anglian herring season in 1912, she made the 250-mile trip to Yarmouth in 31 hours, using only 18 gallons of paraffin. The success of the *Maggie Jane's* encouraged other skippers to follow her example. In 1910, forty-five of the 910 drifters at Great Yarmouth for the Home Season were Scots motor boats.

Several fifies were converted to motor power and had their mizzen mast in front of the wheelhouse, such as with the *Spes Bona* BK 123, seen in the photograph below. Others with this arrangement included the *Elspeth Smith* KY 38, *Refuge* KY 306 and *Celandine* ML 271 of St Monans. As the massive sailing mast was no longer needed, the scottle for the foremast was usually planked over, and the lighter new mast was put in a tabernacle.

Many motor fifies were built, including the *Thrive* BF 267, seen in the drawing. Built in 1927 by Nobles of Fraserburgh for Skipper W. Ward of Portnockie to replace a steam drifter, she was 52 feet 6 inches long with a 16-foot beam and was heavily constructed of 1⅜-inch planking on oak frames. The *Thrive* was powered by a 60/70 hp Kelvin petrol/paraffin engine, which was forward. Some fishermen liked the engine forward as it gave a quiet cabin aft; the disadvantage was that the exhaust came through the side of the boat, so they suffered its fumes when working on deck. Behind the engine was the netroom, which filled up the middle of the boat back to the wheelhouse. As usual, there were wings on either side of the netroom to carry her catch. Forward of the engine was her rope room, where the bush rope for her drift nets was coiled. Her cabin was aft and she was steered by a worm gear. When the boats were moored alongside each other, crews crossing over to their boats were liable to get coated in grease from the worm steering gear.

The 49-foot motor fifie *Milky Way* BF 336 was built by James Noble of Fraserburgh for Skipper John Anderson of Macduff as a drifter and long-liner in 1934. In 1936 she was bought by Eyemouth skipper Peter Maltman, who replaced her 44 hp Kelvin diesel with a 66 hp. She was adapted for seining and was later fitted with a trawl winch for prawn trawling. After a thirty-two-year career at Eyemouth, she was bought by Jimmy and Magnus Sinclair of Scalloway, Shetland, in 1968, being registered as LK 106 and adapted for scallop dredging. In 1973, when nearing her fortieth year, she was completely refitted with a new wheelhouse and galley, a new 112 hp Kelvin engine and her cabin was moved aft. The fore part of her hold was converted to rope bins for her seine ropes; these were an alternative to reels and enabled her deck to be free of coils. Readily recognisable by her yellow colour scheme, she fished as an inshore seiner until 2000.

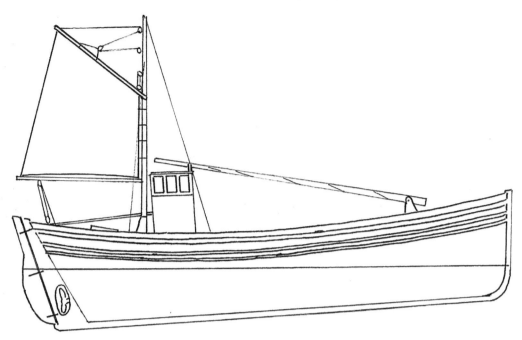

Motor fifie.

The Eyemouth motor fifie *Spes Bona* BK 123 is well down in the water with a good shot of herring. Her crew are busy unmeshing fish. Her net buffs are aft, by the wheelhouse.

British Motor Fishing Vessels

SEINE YAWL

After the First World War, the herring markets in Russia and Germany collapsed and the Danish seiners, which the Scots drifter men encountered at Grimsby, seemed to show a way forward. All along the East Coast were fleets of steam drifters with expenses to pay and much-reduced earning opportunities. Many fishing families also owned small yawls for working lines or creels. These boats had helped out when the steam drifters were laid up between herring seasons; they could be readily adapted for seining and had only a fraction of the running expenses of the drifters. The early seine yawls, like the *Victory* LH 98 and *Victual* LH 327 of Port Seton, had a simple winch, which was belt driven from the engine and facing fore and aft, as to begin with the boats were anchor seining and hauling their ropes and net over the side. With unprotected drives, some of these early winches were dangerous, and several fishermen were caught by their oilskins and injured.

By the end of 1920 there were twenty-seven yawls seining from Aberdeen, but they were soon in conflict with local line boats and creelers, whose crews saw their living being jeopardised by the seiners. The law banned trawling in Scotland's coastal waters, but was unclear about seining. Several seiner skippers were prosecuted, but this was sorted out in 1921, when the law allowed seine boats of under 40 feet in length to fish in coastal waters. Many boats were built at just under 40 feet long to take advantage of this law. So while steam drifters, fifies and zulus were adapted for seining, there was still a place for the yawls on the inshore fishing grounds.

Typical was the *Fidelity* BF 92, built for J. & A. Mair of Cullen. Of 24 net tonnes, she was 39.9 feet long and powered by a 60 hp Kelvin petrol/paraffin engine. The little *Earn* BF 16, built by the Stephen boatyard at Banff for Skipper A. Watson of Whitehills in 1934, was 39.9 feet long with a 12.16-foot beam and was also powered by a Kelvin. She was later owned in Portsoy and Macduff. She was originally a basic little boat with an open deck, but she was later updated with a wheelhouse, a mizzen mast and derrick for getting up the net and a foremast and derrick for landing her catch.

Very early in the story of Scottish seining, they changed from anchor seining to fly dragging. The anchor seiner hauls her gear at anchor and the ropes come in over the side; however, with fly dragging, the seiner does not anchor. When she has set out her ropes and net, she returns to her dan and picks up the end of her first rope. She then tows her net while at the same time hauling her ropes over the stern. Fly dragging is efficient but burns more fuel than anchor seining. The folklore about the invention of fly dragging is that, if the boat was fishing illegally, as sometimes happened, with no anchor to haul, she could more quickly escape from the Fishery Cruiser.

At least one ex-seine yawl is still going strong. The 30-foot *Robina Inglis* LH 361 was built in 1923 by Allan & Brown of Newhaven for Tam Wilson who named her for his wife. She was a decked fifie yawl fitted with a Kelvin petrol/paraffin engine and a mast and sail. In the 1930s she was sold to the Dougal family of Eyemouth, who worked her as the seining yawl *Good Hope* BK 116. In the 1940s she returned to Newhaven, where she was restored by David Brand. She was again extensively restored by Berwickshire Maritime Trust in 2012–13 and renamed *Good Hope*. (The words yawl and yole seem to be interchangeable in Scotland, but mean very much the same.)

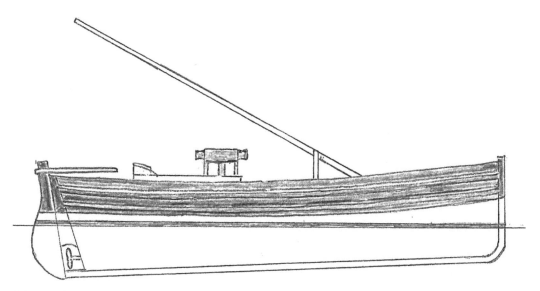

Seine yawl.

Seine yawl *Briar* ML 10, Skipper John Bowman, Pittenweem. Her winch and coiler are under the lowered mast, and her deck is full of coils of seine net rope. The little *Briar* was still at work in the 1950s. In the foreground is the yawl *Athene* KY 240.

British Motor Fishing Vessels

1920S PIONEERS, THE CUTTY SARK FR 334 AND MARIGOLD INS 234

The 45-foot *Cutty Sark* FR 334, completed for Skipper J. Smith by J. & G. Forbes of Sandhaven in early 1928, was a multipurpose cruiser-sterned boat, which set the style for over thirty similar craft – some for Eyemouth and others for the harbours on the Moray Firth. She was designed with accommodation for five crew to go herring drifting, seining or lining. The capstan for her drift net bush rope was forward, and was driven by its own seperate engine. Her seine winch and coiler were behind the wheelhouse. The idea was that she could adapt to any fishery with the minimum amount of bother. Her engine, a Petter semi-diesel with cartridge starting, was in a forward engine room. On her trial trip of 60 miles, her fuel bill was calculated to be 2½d per mile at a speed of 8 knots. She also carried a full set of sails. It was usual for motor boats to be fitted with worm steering gear, a screw thread going from the back of the wheelhouse to the rudder head, but the *Cutty Sark* was fitted with chains from her wheel to the quadrant on her rudder head. This soon became standard for motor boats, as it was for the steam drifters. Her rudder quadrant was covered by a platform, from which her seine net was shot, and this also became standard practice in cruiser-sterned boats. At a time when motor fifies with 'outdoor' rudders were still being built, the stylish *Cutty Sark* popularised the cruiser stern. The miners' strike of 1926 caused the steam drifter fleet to be laid up for a lack of bunkers. This was another incentive to build motor boats like the *Cutty Sark*.

Skipper John Campbell of Lossiemouth was a very experienced fishermen, who had skippered a sailing fishing boat aged only eighteen, and became a partner in the steam drifter *Glengynack*. After service in the First World War, he went trawling in his steam drifter *Marigold*. In 1921 the Lossiemouth fishermen began to adopt the seine net and Skipper Campbell adapted his drifter for seining, but he very soon became convinced that steam drifters were uneconomic and that purpose-built motor boats were needed. His carefully worked out plans for a motor seiner were widely criticised, but he soon proved the doubters wrong.

John Campbell's motor fifie *Marigold* INS 234, the first purpose-built seiner in Scotland, was constructed by William Wood & Sons of Lossiemouth in 1927. The accommodation was forward and the engine aft. Engines were much larger and heavier in those days. Only in bigger boats was there room to have the cabin aft with the engine in front of it, which is now the standard arrangement. The *Marigold* was powered by a three-cylinder 36 hp Gardner semi-diesel engine. Her seine winch and coiler, built by Macduff Engineering and driven by a chain from the engine, were in front of the wheelhouse. This arrangement was later modified. She was 50 feet long with a 16-foot beam and had a crew of four men and a boy. There was one mast forward with a derrick for landing the catch. Unlike most boats, there was no netroom or mizzen mast and sail as her skipper had no intention of going herring drifting. Within a short time the *Marigold* proved herself a profitable boat.

Her example was followed by John Campbell's friend James Macleod, who built the very similar *Briar* INS 420. Lossiemouth became one of the top seining ports in Scotland.

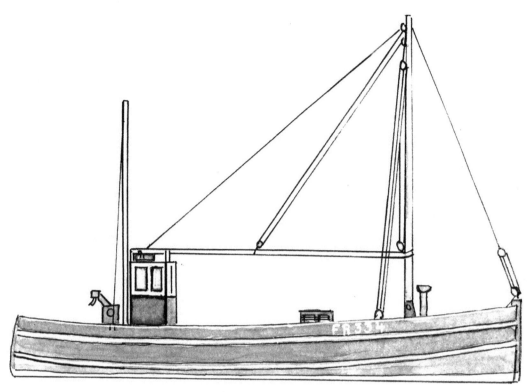

Cutty Sark FR 334.

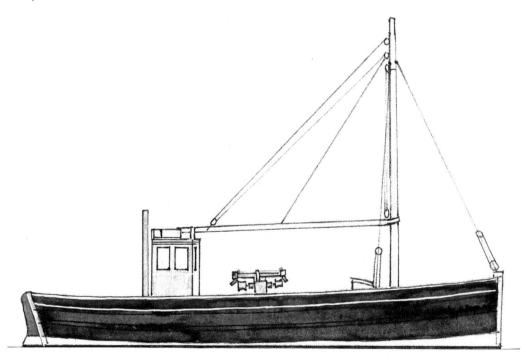

Marigold INS 234.

British Motor Fishing Vessels

SEINERS

The fifie seiner *Sunshine* AH 65 at Arbroath. Her seine ropes are coiled forward of her winch. She is steered by worm gear, with a screw thread from the wheelhouse to the rudder head.

Seiners at Macduff in the 1930s: *Speedwell* BF 303, built locally in 1936, and the *Catherine Margaret* BF 310.

East Coast seiners at Oban: *Undaunted* LH 5 of Granton, *Moray Lass* INS 104 and *Fair Morn* INS 195 of Hopeman.

Adoration KY 187 of St Monans was built at the local James N. Miller yard. She was a classic wooden Scots seiner of the 1950s.

British Motor Fishing Vessels

CRUISER-STERNED MOTOR DRIFTER

Although the market for herring took many years to recover from the effects of the First World War, some owners began to invest in big motor drifters, which were capable of following the shoals around the coasts like the steam drifters. Some of the earliest, such as the *Princess Arthur* BF 386, built in 1920, and the *Cissie* BCK 276, built at Eyemouth in 1928, had counter sterns, but the cruiser stern soon became more usual.

The *Gleanaway* KY 40 was built in 1930 by Forbes of Sandhaven for Provost Carstairs and J. & L. Watson of Cellardyke. She worked from Anstruther. Constructed of pitch pine on oak frames, she was 76 feet long with a beam of 18 feet 9 inches and was powered by a four-cylinder 140 hp diesel. She worked seventy-two herring nets, each of which was 36 yards long and 16 yards deep, her whole fleet stretching for 1½ miles. The *Gleanaway* was sold to South Africa in 1937.

Built for Skipper William Ritchie of Rosehearty by Forbes of Sandhaven in 1931, the *Efficient* FR 242 was 75 feet long, powered by a Petter Atomic diesel engine of 160 hp. In 1937 the *Efficient* was sold to William Stevenson of Newlyn for £393 15s. Initially she worked as a liner and drifter but in 1938 she was converted for trawling. Requisitioned by the Navy in 1941, she returned to Newlyn after the war and was renamed *Excellent* PZ 513 after a lugger once owned by the Stevenson family. After a long career as a trawler and several refits, she was converted to a netter in the 1990s. She was still afloat in 2016, laid up by Newlyn North Pier.

The first cruiser-sterned drifter built by G. Thompson of Buckie was the 80-foot *Girl Helen* BCK 152, which was built for James and John Alex Flett of Findochty in 1934. She was powered by a 220 hp National engine that could be controlled from the wheelhouse – an innovation in 1934. The *Girl Helen* was lost at Dunkirk in 1940.

The *Girl Helen* was followed from the same builders by the shorter but beamier *Poppy* BCK 41, which was 77 feet in length with an 18-foot 6-inch beam and was powered by a 120 hp National engine. Built at a cost of about £3,000, she grossed £495 at her first Great Yarmouth herring season with expenses of £194. In 1948 she was sold to Peterhead and was renamed *Carntoul* PD 394. In 1956 she was bought by Skipper Donald Anderson and renamed *Glenugie II*; sold again in 1966 to Skipper George Nicol, she became the prawn trawler *Ugiebrae*. Updated throughout her career, she was then equipped with a Gardner 152 hp diesel, a radar, a VHV radio and modern navigational equipment. In 1970 she was sold to Swansea owners and was re-registered there.

As economic conditions gradually improved, there was further investment in cruiser-sterned drifters. In 1933 the Peterhead drifter *Caledonia* PD 160 was built at Macduff and the 35-ton *Narinia* INS 133 was built for Nairn owners. In 1936 the hefty 60-ton *Comfort* FR 965, which was powered by a Kelvin six-cylinder 132 hp diesel, was launched for Skipper B. Noble of Fraserburgh. She too was lost at Dunkirk in May 1940. Forbes of Sandhaven built the 65-foot *Three Bells* BCK 114 for Skipper A. Smith of Findochty in 1937. Sold to Peterhead in 1947, she had several owners before being lost south of Wick in 2000. Forbes also completed the *Helen West* BF 363 in the same year.

By the late 1930s many fine cruiser-sterned drifters had been built for the Scots East Coast ports. The trend was to continue after the Second World War until the 1960s.

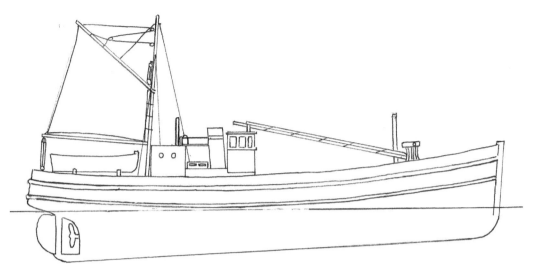

Cruiser-sterned motor drifter.

Skipper William Ritchie's 75-foot motor drifter *Efficient* FR 242 at Lowestoft for the East Anglian herring season. (Photograph Port of Lowestoft Research Society)

British Motor Fishing Vessels

75-FOOT MFV DRIFTER

After the Second World War, the Scots herring fishermen found the 75-foot ex-Admiralty MFVs ideal for conversion to drifters. The over-large deckhouse and funnel were usually reduced or replaced; a netroom was fitted and a capstan was installed forward for hauling the bush rope. Many of these craft had profitable careers, three of them winning the coveted Prunier Trophy while fishing the East Anglian herring season from Lowestoft or Great Yarmouth.

Shetland's 75-foot MFV drifters included the *Southern Cross* LK 39, *Ocean Reaper* LK 64, *Cornucopia* LK 470, *Margaret Reid* LK 440 and *Jessie Sinclair* LK 272, which won the 1954 Prunier Trophy under Skipper Robert Williamson with a shot of 272 crans from the Smith's Knoll grounds landed at Lowestoft on 12 October. The *Jessie Sinclair* was built by Walter Reekie at Anstruther as MFV 1166 and was bought for fishing in 1948. Her engine was the usual 160 hp Lister Blackstone diesel. The 1954 autumn season saw some good catches but was plagued by storms, which damaged nets and kept the fleet in port. The *Jessie Sinclair* was the only Shetland drifter to win the Prunier Trophy. The runner up was the steam drifter *Wilson Line* KY322, with a 270 cran landing at Yarmouth on 7 November.

One of the 1956 trophy winners was the Fraserburgh drifter *Stephens* FR 156, which landed 215 crans on 20 October under Skipper F. Stephen. Other good landings on the same day were from the Fraserburgh boats *Dayspring* FR 125, with 212¼ crans, *Golden Harvest* FR 337, with 207¾ crans, and *Gleaner* FR291, with 201 crans. 1956 was unusual as there were two trophy winners, each landing with 215 crans. Aside from the *Stephens* FR 156, the other was the Lowestoft steam drifter *Silver Crest* LT 46, which landed her winning shot at Lowestoft on 22 October. The 1956 season was generally a very poor one and several Scots boats left for home in the middle of November.

The *Stephens* was built by Forbes of Sandhaven in 1945 as MFV 1188. Sold for fishing in 1947, she was registered in Fraserburgh as *Wests* FR 156. Acquired by the Stephen family, she was renamed *Stephens* in 1954.

The 1960 Prunier Trophy was won by another Fraserburgh 75-footer, the ex-MFV *Silver Harvest* FR 178 with a landing of 187 crans, which sold for more than £1,530 at Yarmouth on 17 November. The *Silver Harvest* was built by John Morris at Fareham in 1945 as MFV 1149 and was and sold by the Admiralty for fishing in 1947. After a useful career at Fraserburgh, she was sold to Ireland, and was registered as D 171, retaining her original name.

Among the many other 75-foot MFVs converted for drifting and seining in Scotland were the *Corona* BCK 27, which was built as MFV 28 by Frank Curtis at Looe in 1942, the *Harvest Gleaner* BCK 120 (originally MFV 1154), from Herd & Mackenzie at Buckie in 1946, the *Nautilus* BCK 122, from Walter Reekie of Anstruther as MFV 1196 in 1946, the *Sea Wave* BCK 124, from Humphrey & Smith of Grimsby as MFV 1180 in 1945, the *Brig o' Nelson* BCK 134, from the same builders as MFV 1099, and the *Easter Morn* BF 76 (formerly MFV 1213), which was from James Noble of Fraserburgh.

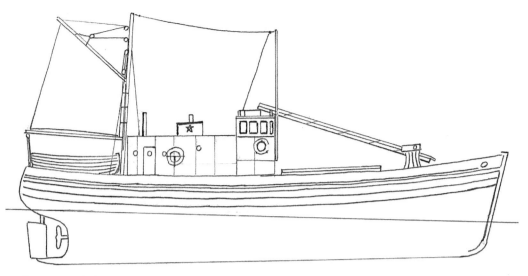

75-foot MFV drifter.

Skipper A. Watt's Fraserburgh drifter *Resplendent* FR 162. She retains her high MFV wheelhouse and funnel. Most Scots fishermen replaced these with more elegant structures. Her nets have been hauled up on deck over the two rollers on either side of her netroom, and two of her crew are mending them. Her net buffs are beside the wheelhouse. Forward she has two of the traditional long larch fenders while the more usual rubber tyres can be seen aft. Beside her small boat are the baskets which will be used to land her catch. Forward, one of her crew is getting ready to hoist up the foremast for unloading her herring.

British Motor Fishing Vessels

50-FOOT MFV

In his *Shetland's Fishing Vessels*, J. R. Nicolson commented on the 50-foot MFV seiner *Silver Cloud* LK 317 of Scalloway: 'A smaller type of MFV was the 50-foot-long *Silver Cloud*, reckoned by many to be one of the finest boats of her size ever to fish from Shetland.'

The first orders for 50-foot MFVs were made in 1942. The boats were 50 feet long and 45 feet between perpendiculars, with a 15-foot beam, an inside depth of 8 foot 1 inch, a draught forward of 3 feet 3 inches and were 4 feet 10 inches aft. They were designed on the lines of the Scots ring netters with the engine aft, the hold amidships and the cabin forward. The big problem for fishermen buyers after the war was that many were powered by Chrysler eight-cylinder 85 hp petrol engines because of the lack of suitable diesels, but these were not suitable for fishing boats.

In October 1945 Pearsons of Hull completed MFV 990. Sold in February 1949 to John Ralph & Partners of Hopeman, she was converted to the ring netter *Fragrance* INS 259. As a ringer, she paired with the *Ardent* INS 326, *Flourish* INS Z53 and *Triumph* INS 146. In 1971 she was sold to Fraserburgh owners Alexander Wiseman & Partners and renamed *Silver Leaf* FR 93. Another seiner was MFV 991, which was also completed by Pearsons in 1945. Sold to William Smith of Buckie in 1946, she was registered as *Bramble* BCK 174. Her Chrysler petrol engine was replaced by a Gardner 95 hp diesel. In 1947 she was sold to Ayr on the West Coast and was renamed *Intrepid* BA 277. In 1960 she returned to the East Coast to Gourdon, which is well known as one of the last small lining ports, where she was registered as ME 68.

The Harbour Shipwright Company of Bridlington completed MFV 968 in 1946. The following year she was bought by J. Addison of Cullen and was registered as *Addisons* BF 68. In 1961 the seiner was bought by Alexander Cowie of Fraserburgh and renamed *Trust* FR 252. In 1980 she was sold to the West Coast of Scotland and five years later to Irish owner Eugene O'Neill of Baltimore, County Cork, and was registered as SO 804.

Other 50-foot Scots seiners included the *Tyro* BF 320 (ex-MFV 967, built in Bridlington 1945), *Coral Strand* BF 342 (ex-MFV 956, built at Kilkeel 1945), *Vesper II* KY 125 (ex-MFV 798, built by Fairmile at Cobham 1945) and *Mizpah* INS 99 (ex-MFV 790, built at Horning, Norfolk, in 1946).

Several former 50-foot MFVs fished in Cornwall. The *Renovelle* PZ 169 was another 50-footer built by Pearsons. When fitted out for the Madron family by Raymond Peake, her cabin was moved aft. The *Renovelle* regularly went lining west of Bishop Rock and a forward cabin would not have been popular. Another former MFV was Skipper Donald Turtle's *Bonny Mary* PZ 57, which had a career long-lining from Newlyn.

In the 1970s, netting for crawfish took off in Cornwall. This developed into deep water netting for pollack, hake, turbot, cod and ling. Among the many netters bought in Cornwall were the 50-foot MFVs *Confide* PZ 741, which was built as MFV 772, and *Kim Bill* PZ 253. Both were constructed in Poole.

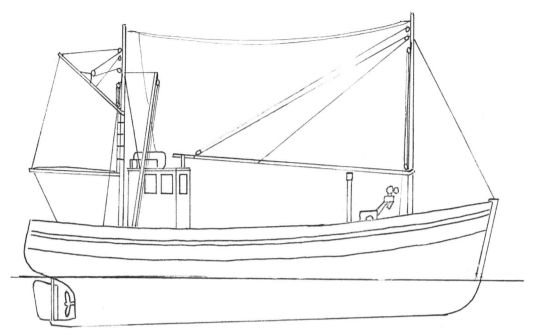

50-foot MFV.

The Mousehole liner/drifter *Renovelle* PZ 169 being fitted out at Newlyn by boatbuilder Raymond Peake *c.* 1955. Owned by the Madron family, the *Renovelle* went long-lining in the spring and early summer. This was followed by pilchard drifting. For several years she worked a ring net at the autumn pilchard season and for herring at Dunmore East in Ireland. The *Renovelle* and the *Sweet Promise* SS 95 of St Ives made a success of ring netting.

British Motor Fishing Vessels

FIFER CREELER

Wherever there are suitable rocky fishing grounds, there are likely to be fleets of potters or creelers. The East Coast of Scotland was no exception and the small ports of Fife, such as Crail and St Andrews, had their iconic fifer creel yawls. A typical fifer was double-ended with an outside rudder and the usual two rubbing strakes on either side, making it a miniature version of the fifies already described. Her hauler and its rope roller were aft, where she worked her lines of creels. The side of the boat was fitted with protective strips to save it from chafe by the creels as they came aboard. Her mast was right in the bows, usually being lowered down in its mast crutch. In the past the mast carried a lug sail, but by the mid-twentieth century its main job was to carry the lights. Spare gear like the oars and boat hook were often stowed along the mast. The working part of the boat was surrounded by narrow side decks with a coaming to keep out any water that came aboard and there was a foredeck and aft locker.

Since their earnings were likely to be more modest, many of the small yawls were motorised long after the big fifies and zulus. These small craft were used initially for creels and lines, but from the 1920s seining began to replace lining. One of the earliest motor yawls was the 27-foot-long *Vanguard* 19 AH, built by Millers of St Monans in 1910 for Arbroath and fitted with a 7½ hp Gardner engine. She was followed the next year by another Arbroath-registered yawl, the *Restless Ocean* AH 27, also from Millers, which was fitted with a 5 hp Gardner engine that drove a three-bladed propeller to give a speed of 7 knots. The *Restless Ocean* had a length of 28 feet and a 10-foot beam. As in other small craft, the highly reliable Kelvin 3½ hp and 7 hp petrol/paraffin engines were popular until the 1960s, when small diesels became available.

The James N. Miller boatyard at St Monans was well known for building fifer creel boats. In 1950 Millers built the 28-foot-long *Marean* KY 120 for Skipper Blackery of Crail. She was carvel-built of larch planks on oak timbers and driven by a 22 hp Kelvin J2 engine. Creelers built for the Meldrum family of Crail included the *Sovereign* KY 126 and *Shirley II*, which were completed in 1956. Both were fitted with Kelvin 10 hp P2 engines. In 1959 Millers built the *Lily II* KY 210. In the 1960s they advertised three versions: their 20-foot-long fifer had a 6-foot 6-inch beam, a draught of 2 feet 3 inches and was powered by a 7 hp Kelvin; the 26-foot version had an 8-foot 6-inch beam, a draught of 2 feet 9 inches and was driven by a 15 hp Kelvin; and the largest was 30 feet long with a 10-foot beam, a 3-foot 6-inch draught and a 30 hp Kelvin engine.

Traditional Fife lobster creels were made of bows of ash wood bent over a wooden base and covered with net, and the entrance was formed of a funnel-shaped net. Sometimes the ash bows were steamed into shape.

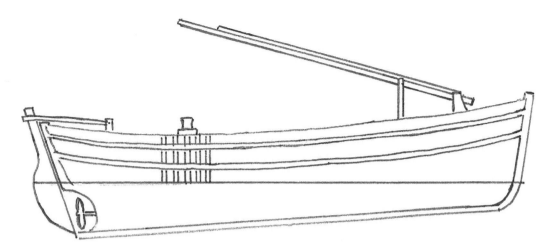

Above: Fifer creeler.

Right: Fifer creelers at Crail in 1968: *Aspire* KY 266, *Sovereign II* KY 333, *Comely* KY 175 and *Be in Time* II KY 3. The *Sovereign II* has a radar reflector on top of her mast. Though only 28 feet long, she has a cruiser stern. In 1968 she hit the headlines by accidentally catching a 7-foot-long leatherback turtle and taking it back to Crail. The unfortunate creature was entangled in the boat's creel ropes. The other creelers have fifie sterns with the rudder outdoors. The *Be in Time II* and *Aspire* have just returned from hauling their creels.

British Motor Fishing Vessels

RIPPER YOLE

In 1948 the boatbuilders Tommy Summers, Bill Duthie and George McLeman, who had all served their time with the local James Noble boatyard, set up their own boatbuilding company, known as Thomas Summers & Company, at Fraserburgh. Their first boat was the 33-foot-long ripper yole *Fisher Boys*, which was built for George Duthie. She was an advance on previous cod hand-liners as she had a cruiser stern instead of the traditional fifie stern with an outside rudder, and she was fitted with a proper wheelhouse to shelter the helmsman. The wheelhouse, with its floor set down below deck level, was carefully devised so that the skipper could handle the boat and work his own line. Such was the success of the *Fisher Boys* that she was the first of a whole series of twenty-nine similar craft built by Tommy Summers & Co., including the *Daisy* FR 78, *Jeannie West* FR 146, *Gracious* FR 167, *Lenten Rose*, *Hopeful*, *Valiant* FR 231, *Grateful* FR 270, *Fair Morn*, *Kittiwake* and their final yole *Daybreak*. There was a big demand for these boats in the 1950s and '60s, and often two or three were being built alongside each other by the Summers yard. Many apprentice boatbuilders learned their trade building these neat little craft. Several ripper yawls are still going strong from other ports, though they have been adapted for different fisheries, sometimes with the wheelhouse re-sited forward. At nearly sixty years of age the little green *Lenten Rose* is still at work as the miniature trawler FY 43 at Mevagissey. The *Hopeful* also came to Cornwall and worked crawfish nets from St Ives in the 1970s.

The ripper line was a hand-line fastened to a 2½ lb weight called a sproule, from which hung the ripper – a short lead with about four hooks attached. Later, feathered hooks were attached to the line between the sproule and ripper. Although various refinements have been tried over the years, the original ripper gear is still effective. The main changes to the gear resulted from the arrival of artificial fibre lines and monofilament. Typically, a yole had three crew who would land up to six boxes of cod per trip. Freshly caught the same day, their catches were of high quality and fetched good prices. The yoles fished on hard ground, which could not be worked by seiners or trawlers. Spring was the main season for cod rippering; during the summer the yoles went lobster creeling or feathering for mackerel, and in winter some of them worked small lines.

Although designed for hand-lining, they had very low bulwarks and their crews must have needed good sea legs to stay aboard. Originally fitted with petrol/paraffin engines, by the 1960s many had Kelvin or Gardner diesels. The engine was aft, the hold amidships and the cabin was forward. Like the rest of the Scots fleet, the yoles were beautifully maintained, until the 1960s most of them sticking to the traditional black hull with a yellow line at deck level, a white waterline, a varnished and grained wheelhouse and white spars. Exceptions were the green-painted *Gracious* and the bright red *Hopeful*. Their regular berth was in Fraserburgh Balaclava Harbour, for many years with the veteran zulu *Violet* FR 451.

Fraserburgh had the largest fleet of ripper yawls but many other Scots harbours worked similar fisheries, as did Seahouses and Beadnell in Northumberland.

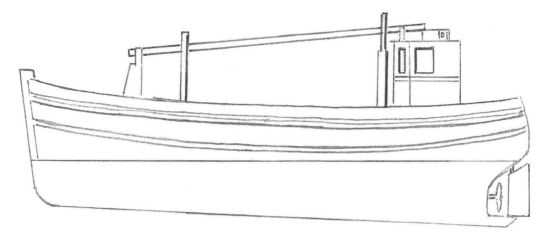

Ripper yole.

Ripper yoles *Daisy* FR 78 and *Jeannie West* FR 146 at Fraserburgh Balaclava Harbour in 1968.

British Motor Fishing Vessels

STEEL STERN TRAWLER

Sailing smacks worked their beam trawls off the side. When the otter trawl arrived, it was worked in the same way, off the side of the boat. These trawlers are now called side winders. Even boats built with the wheelhouse forward, like the Scots-built *Green Cormorant* PZ 53, worked their trawls off the side in the usual way.

The first British factory stern trawler, Christian Salvesen's *Fairtry* LH 8, began work in 1955. Her trawl was worked from a stern ramp. She was followed by a whole series of large stern trawlers for Hull, but inshore and near-water side winders continued to be built in the 1960s and '70s.

In 1961 pioneering Skipper Charles Scott of Fleetwood built the wooden 44.8-foot stern trawler *Onward Fisher* at J. & J. Harrison of Amble and worked the Clyde, Isle of Man, Solway Firth and Irish coast grounds. Skipper Scott was soon convinced of the merits of trawling over the stern. The gear was shot and hauled much quicker than in a side trawler, the work was safer and there was no hand pulling; the winch did the work.

In 1965 Forbes of Sandhaven built the wooden transom-sterned stern trawler/seiner *Constellation* FR 394, which is believed to have been the first of her kind in Scotland. At that time all Scots boats were built with cruiser sterns. Skipper J. Buchan's *Constellation* had her wheelhouse and winch forward, a swinging gantry aft and a goalpost mast amidships. She was 65 feet long with a 20-foot beam and an 8-foot 6-inch draught, and was powered by a Kelvin 240 hp T8 engine. Her wheelhouse equipment included an Elac echo sounder and fish lupe, a Woodsons radio telephone and DF set, and a Decca Navigator and Track Plotter. It was reported that other yards had designs for stern trawlers, but skippers were reluctant to order them while they could make a living with their conventional boats.

In 1966 Herd & Mackenzie of Buckie built the steel stern trawler *Golden Strand* BF 403. In 1968 she was followed by the similar *Janeen* BCK 5 for local owners Smith & Jappy. The *Janeen* was 50 feet long with a 15.5-foot beam and an 8.5-foot depth, and was driven by a Caterpillar 210 hp diesel. Her wheelhouse equipment was similar to the *Constellation*'s, but she also had a Kelvin Hughes radar. Although stern trawlers were becoming more popular, it was still unusual for inshore boats to be built of steel. The *Janeen* had a very long career from Buckie, Shetland, Aberdeen, Kilkeel and Leigh-on-Sea. Bought from Leigh in 2013, she was converted to a crabber at Newlyn and was still at work in 2016 as a mussel farm boat in East Cornwall.

In 1968 Campbeltown Shipyard was established to build steel fishing boats. Among its first builds were the small stern trawlers *Crimson Arrow* CN 30 for local skipper James Macdonald, the *St Adrian* KY 245 for Skipper David Tod of Anstruther and the *Terra Nova* A 219 for Aberdeen. The *St Adrian* was fitted with the then innovative net drum and hydraulic winch. It was these boats that put the small steel stern trawler on the map. Steel was stronger than wood and could better withstand the strains imposed by trawling and scallop dredging, which were becoming more popular.

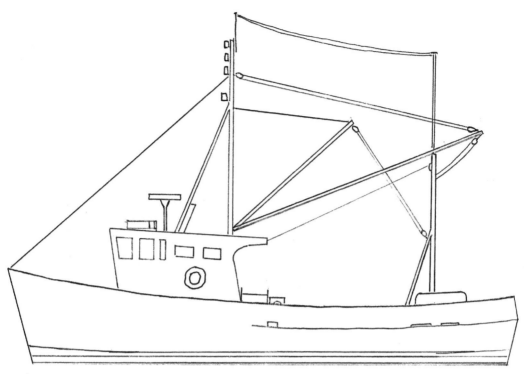

Steel stern trawler.

The wooden 65-foot *Constellation* FR 294, built for Skipper J. Buchan of Fraserburgh by Forbes of Sandhaven in 1965, is believed to have been the first transom-sterned stern trawler in Scotland.

British Motor Fishing Vessels

CAMPBELTOWN 80-FOOT SEINER

In 1972 the first 80-foot steel seiner *Argosy* INS 79 was launched from Campbeltown shipyard, which had already won its reputation with 50-foot stern trawlers. The *Argosy*'s design was developed with the help of Skipper Willie Campbell of Elgin, basing it on his wooden seiner *Ajax*. In contrast to the chine-hulled stern trawlers, the Campbeltown 80-footer was a round-bilged vessel. The first two orders for these boats were from the highly respected seine net skippers William and Andrew Campbell. They were followed by two orders from Buckie, one from Hopeman and one from Fraserburgh.

The *Argosy* was 79.95 feet long overall with a 22-foot beam, an 11-foot depth and a draught aft of 10 feet. Her beam was carried well aft to give plenty of room for working her seine net. She was powered by a Caterpillar 480 hp diesel, giving a speed of 10 knots, and had a Lister auxiliary engine. The plating of her all-welded hull was carefully treated to avoid corrosion. She was fitted with a Jensen seine and trawl winch with a Beccles coiler forward, where there were also two fish washers. There was a crane aft and a Carron power block for hauling her net. Her wheelhouse was well equipped with an autopilot, Kelvin Hughes radar, echo sounder, a watch alarm, radio, VHF radio and Decca Navigator. Her fishroom was designed for boxing at sea and was insulated with Solarfoam. An ice-making machine was to be fitted under her whaleback. Her centrally heated cabin aft was fitted out for eight crew; meanwhile, the skipper's cabin was in the deckhouse, where there was also a mess room and a well-equipped galley fitted with an oil-fired cooker and oven.

The *Argosy* began work from Peterhead. In contrast with the sputnik trawlers, which took several years and various adaptations to prove themselves, the Campbeltown 80-footers were an immediate success. They were soon among the highest-earning fishing vessels in Scotland. Among the record breakers were the *Argosy* INS 79, *Ajax* INS 82, *Kestrel* INS 121, *Emma Thomson* INS 100, *Xmas Rose* FR125 and *Argonaut IV* KY 157. Throughout the 1970s and '80s they were modified with gutting shelters, seine net rope reels and shelter decks. A fully shelter-decked Campbeltown 80-footer was an incredibly seaworthy vessel that could carry on fishing in very poor weather, while at the same time keeping her crew safe. So successful were they that other variants of the design were built: 75 feet (*Mary Croan* INS 231), 85 feet (*Spes Melior* PD 379 and *Boy Andrew* WK 171) and 88 feet (*Ardent* INS 127).

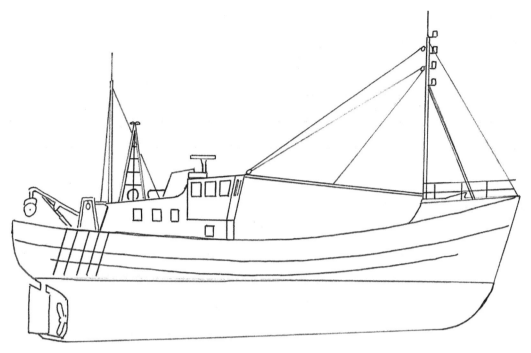

Campbeltown 80-foot seiner.

Skipper Damien Turner's Campbeltown 80-foot seiner *Roise Catriona* T 100 of Castletownbere, Ireland, in 2010. She was built as the *Argyll* INS 217 for Skipper Brian Walker of Burghead & Partners in 1982, powered by a Deutz 554 hp engine, and fished from Peterhead. Built with a whaleback and gutting shelter, she was subsequently fitted with a three-quarter-length shelter deck. The *Roise Catriona* was still fishing in 2016.

British Motor Fishing Vessels

MOTOR LOCH FYNE SKIFF

The ring net developed from the idea of surrounding a shoal of herring with drift nets and dragging it to the nearest shore like a beach seine. By the mid-nineteenth century the ring net had arrived. It had floats at the top and leads along the bottom and was worked by two boats, usually close inshore. On spotting a likely shoal, one boat would shoot her net around it. Her neighbour boat then picked up the end of the net and both boats would tow the net around the shoal until they met. The net was hauled by its owner boat, with both crews joining in. The bottom of the net was attached at intervals to the spring rope, which was hauled in from both ends, lifting up the bottom of the net and the fish in it. At the same time, the net itself was hauled in from both ends until they got to the fine meshed bunt in the middle. The herring were now swimming about in the net alongside, from where they were dipped out in baskets. Compared with working a fleet of drift nets, which took several hours, ring netting was a speedy business and several shots could be made in a night.

A special kind of boat, the Loch Fyne skiff, was developed in the Clyde ports – Carradale, Tarbert, Campbeltown, Dunure, Maidens and Girvan – to work the ring net. Up to 35 feet long, the skiff was shallow forward and deep draughted aft with a curved stem and sharply sloping pointed stern. These were qualities that made her very manoeuvrable for working the net. She carried a high-peaked standing lug sail on a raked mast right forward. There was also a jib and bowsprit. The cabin and foredeck took up the forward third of the boat, the rest of which was undecked. Loch Fyne skiffs were usually varnished and well maintained.

The first engine, a 7/9 hp Kelvin petrol/paraffin model, was fitted to a Campbeltown skiff in 1907. It was an immediate success, and within a few years dozens of skiffs were fitted with engines. To begin with, only one boat from each pair was motorised, but it was soon evident that having engines in both boats produced much higher earnings. Engines were fitted aft, with the propeller on the starboard side, since West Coast ring netters worked their gear over the port side. Very soon more powerful engines were installed, and new skiffs were fitted with 13 to 15 hp engines. Engines made a huge impact on the ring netters but, to begin with, their basic design remained largely unaltered.

When the fishermen returned from the First World War, the West Coast herring industry was slow to recover, and they were reluctant to invest in new boats, which were much more expensive. Although Robert Robertson of Campbeltown built the first pair of canoe-sterned ring netters in 1922, traditional Loch Fyne skiffs continued to be built.

In 1923 J. N. Miller of St Monans completed two similar decked 43-foot skiffs, which had been designed by W. G. McBryde of Glasgow. The first, intended for James Gibson of Dunure, was powered by a 40 hp Atlantic engine with the propeller on the starboard side. The mast came through a scottle on the foredeck into the cabin, in the traditional way. Millers built another similar skiff for Duncan Munro of Loch Fyne; this time 40 feet long with a 12-foot 6-inch beam and a 4-foot 6-inch draught. Constructed of pitch pine planks on oak frames, she was driven by two 13/15 hp Kelvins – one on the centre line and one to starboard. Her rudder was cut out for the centre propeller.

In 1926 two other McBryde-designed skiffs were built for Loch Fyne by Wilson Noble of Fraserburgh. Each was 43 feet long with a 13-foot beam and a 4-foot 6-inch draught, powered by a 28 hp Glennifer engine. They carried 21 gallons of fuel and had accommodation for four crew. Although they still carried a sail, there was less reliance on it, and as a result the mast was stepped in a tabernacle on the foredeck.

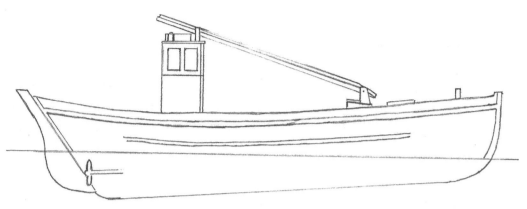

Motor Loch Fyne skiff.

Loch Fyne skiff at Portree, Skye.

CANOE-STERNED RINGER

In 1922 James N. Miller of St Monans built two revolutionary ring netters for Robert Robertson of Campbeltown (who had been the first to install a motor in his skiff back in 1907). Designed by W. G. McBryde along the lines of boats Robertson had seen in Norway, the *Falcon* CN 97 and *Frigate Bird* CN 99 were 10 feet longer than any other skiff, fully decked, with canoe sterns, curved stems and rounded forefoots, and were fitted with wheel steering and wheelhouses. They were 50 feet long with a 15-foot beam and a 5-foot draught. They were driven by two Glennifer paraffin engines of 18/22 hp and 25/30 hp. One engine could be controlled from the wheelhouse, which was an advanced feature at the time. The main engine propeller was on the centre line, with the second to starboard. The boats were built of pitch pine on oak frames to the highest standard. They were fitted with 18-foot-high bulwarks, but the foredeck was raised to give a comfortable cabin, and was equipped with a skylight and companion. The boats were fitted with acetylene gas lights as well as a large sail plain of fore and mizzen lugsails to help them along and save fuel. The *Falcon* and *Frigate Bird* were initially criticised, but soon proved themselves. Despite their success, traditional skiffs continued to be built, as described previously.

In 1927 Robert Robertson returned to Millers for another canoe-sterned ring netter – the smaller *Crimson Arrow* CN 208, which was strongly built of pitch pine on double oak frames. The *Crimson Arrow* was 41 feet long with a 12-foot beam and a Kelvin 26/30 hp paraffin engine. She was fitted with a large auxiliary lugsail. A year later Robertson and Short built the slightly bigger 46-foot *Nil Desperandum* CN 222, which had the unusual feature of having a wheelhouse forward, to give a clear deck for working the ring net. However, this was not a success, and it was soon moved aft to the conventional position.

Hauling up the spring rope by hand was gruelling work and various attempts were made to design a satisfactory winch. This arrived in 1929 and was quickly adopted by the whole ringer fleet. Robertson had built the smaller and lower *Crimson Arrow* as, although the *Falcon* and *Frigate Bird* were successful, there was huge labour in emptying their ring nets with stick baskets. This was solved by the introduction of the brailer – a giant version of a child's shrimping net, which dipped the herring out of the net and was hoisted up by the brailer pole, using the winch. The first ringer to use the brailer was the *Mary Sturgeon* in 1933. Like the winch, it was soon in use by the whole fleet.

By the late 1920s the canoe-sterned ringer was accepted and they were soon being built by well-known yards like William Weatherhead of Cockenzie (who built their first, the *Mary Sturgeon* BA 40, in 1926), as well as by Millers and Walter Reekie of St Monans and James Noble of Fraserburgh, whose boats were all readily recognisable and were widely thought to be some of the most beautiful working boats built anywhere.

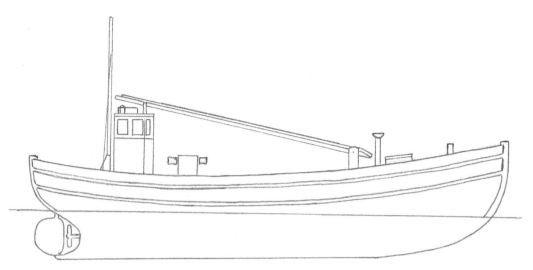

Canoe-sterned ringer. (Drawing above: Kestrel and Kittiwake, R. Robertson, 1930)

Ring netters at Bridlington, most of them from Maidens, Ayrshire, and all built by Weatherhead of Cockenzie (except for the *Marion*): *Maireared* BA 196, built in 1933, *Investors* BA 58, built in 1939, *Margarita* BA 56, built in 1934, *Marion* BA 170, built by James Noble in 1936, *Goodwill* LH 106, built in 1934 from Musselburgh, and *May* BA 20. Most of them have their ring nets and buffs aft. The sloping derricks in front of the wheelhouses are brailer poles used for emptying the net. The *Maireared*'s cabin stove pipe and skylight are in the foreground. The oilskins round the *Investors*' wheelhouse windows give privacy to fishermen using the bucket, as there were no toilets.

British Motor Fishing Vessels

POST-WAR RINGER

The typical post-war ring netter was about 50 feet long with a beam of a little over 15 feet and a 5-foot draught aft. Her diesel engine was of 88 to 100 hp, giving a speed of over 9 knots, and she carried enough fuel for a week's work. The engine room was equipped with a dynamo to provide electricity for the lights, a bilge pump and a drive to the ring net winch in front of the wheelhouse. Her engine room was aft, the fish hold, which could carry about 150 crans of herring, was amidships and the cabin was forward. The cabin was fitted with berths for six, a table, cupboards, lockers and a coal stove for cooking and heating. There was a skylight on the foredeck.

The hull was fitted with two rubbing bands and was designed with a rounded forefoot and balanced rudder to make her manoeuvrable. As ringers came alongside each other every night, they were easily recognised by having had their starboard side lined with rubber tyre fenders. They had a marked sheer as they needed a high bow to keep them dry, but a low freeboard amidships to enable them to work the ring net and empty its catch with the brailer and winch. They had to be unobstructed aft to enable the net to be shot straight over the stern, so there were no bollards or fairleads on the gunwale. There was usually a thwart across the stern, which was used for mooring the boat. For the same reason, the rod and chain steering gear and the rudder quadrant were below the deck.

Fish finding was with the feeling wire – a length of snare wire with a weight attached, kept wound around the end of a fish box when not in use. A skilled man could feel the herring hitting the wire. Radio was adopted as soon as suitable sets became available, and by the 1950s the echo sounder was in use. The wheelhouse was small in order to give as much working room as possible on deck, but larger wheelhouses were soon built to make room for radios, echo sounders and other equipment.

The use of the ring net had also spread to the ports on the Firth of Forth like Fisherrow and Port Seton, as well as to Avoch and Hopeman on the Moray Firth. Portavogie in County Down also had its fleet of varnished ring netters, many of which were built by J. Tyrrell of Arklow, including the *Essie Orr* B 312, *Eleanor Annetta* B 124 and *Oriele*. While Clyde ring netters worked their nets off the port side, the East Coast boats worked theirs to starboard. The ringers annually migrated to the herring fishery in the North Sea and fished from Seahouses, Whitby and Scarborough. They also sailed to the Isle of Man and fished from Peel. The Isle of Man had followed the success of the Scots and built its own ring netters in the 1930s to regenerate its local fisheries, including *Manx Fairy* PL 43, *Manx Beauty* PL 35, *Manx Lad* PL 23 and *Manx Lass* PL 33. The main Manx destination for herring was the kipper business, and once the smokehouses were fully supplied, other catches were run across to Portpatrick, where they usually fetched a much lower price.

In the winter, the most able fishermen from the Clyde ports worked from the Outer Hebrides, running their catches across the Minches to Mallaig, which became the top herring port in Europe. The arrival of the 50-foot Scots ring netters *Arctic Moon* BA 369 and *Elizmor* BA 343 and their highly skilled crews at Dunmore East in 1955 revolutionised the Irish herring fishery, and their example was quickly followed.

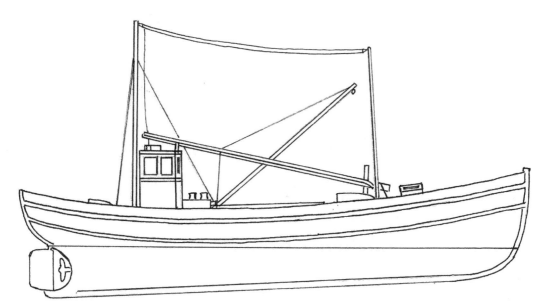

Post-war ringer.

Mallaig ring netters *Spindrift* OB 139 and *Primrose* OB 172 were both built by James Noble of Fraserburgh in 1946. Their ring nets and buffs are aft, on the port side. In front of the wheelhouse, their brailer poles are against the lowered masts. The cabin skylights are open.

British Motor Fishing Vessels

RINGER TRAWLER

During the 1950s and '60s, prawn trawling and scalloping developed. The ring netters adapted to the situation and were fitted with trawl winches so that they could diversify into other fisheries as alternatives to ring netting for herring. These multipurpose ringer/trawlers increased in size and engine power. In 1946 the traditional builders of ring netters were joined by the new boatyard of Alexander Noble of Girvan, which quickly earned a high reputation by building many of the last generation of these craft. By the 1960s they were being built up to 60 feet long with 200 hp engines. With increased power and their role as trawlers, they were made fuller aft and the cruiser stern replaced the canoe stern.

In 1964 Herd & Mackenzie of Buckie built their first ring netter, the hefty 60-foot *Falcon* INS 235, for J. Sutherland & Partners of Hopeman on the Moray Firth. Powered by a Gardner 150 hp diesel, she was equipped with a Reid seine net winch and a Beccles coiler to enable her to work as a seiner. Her wheelhouse equipment included a Kelvin Hughes echo sounder, a Decca Navigator and a Marconi radio telephone.

The 59-foot *Aliped* IX was completed by Alexander Noble for Andrew, Edward & Thomas McCrindle of Girvan in 1964. Powered by a 180 hp Kelvin, she was fitted with the usual ring net winch in front of the wheelhouse and a trawl winch forward for prawn trawling. She was one of a whole series of 'Alipeds' built for the same owners. The following year Nobles built the 60-foot ring netter *True Token* B 600 for David Adair of Portavogie, Northern Ireland. Driven by a 200 hp Gardner, she was also intended for prawn trawling, seine netting or herring drifting. When ringing, the *True Token* neighboured the locally constructed *Glorious* B 428.

The 60-foot ringer/seiner *Ocean Maid* CN 266 was constructed by Fairlie Yacht Yard in Ayrshire for A. & M. Macmillan of Carradale in 1967. Her 365 hp Caterpillar engine gave her a speed of nearly 11 knots. Her fishroom could hold 200 crans and her cabin had accommodation for seven crew.

From the mid-1960s, midwater pair trawling for herring developed and many powerful pair trawlers were built on the East Coast. In 1966 the first British purse seiners – the *Glenugie III* PD 347 and *Princess Anne* LT 740 – were fitted out, and ambitious skippers began to build pursers. Many of the larger ringers were adapted to pair trawl herring. Pursing and pair trawling caught vast quantities of fish, and ring netting declined during the 1970s, with the method last being used in 1978. The pursers and pair trawlers demonstrated their ability to the extent that the herring stock was endangered and limits had to be imposed.

The ringers were very solidly built and many of them continued to earn their keep as prawn trawlers into the twenty-first century. While the ring netter was regarded as a thoroughbred, perhaps the prawner became more of a plodding workhorse.

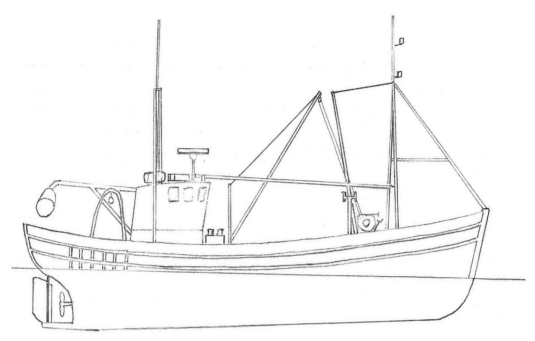

Ringer trawler.

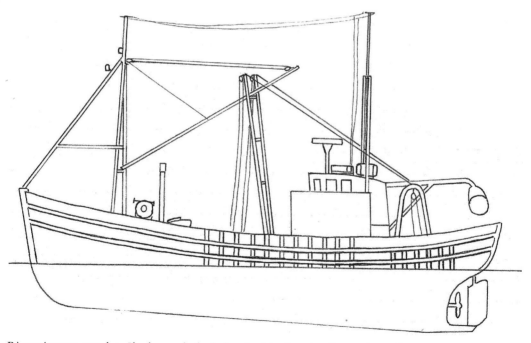

Ringer/prawn trawler. She has a deck shelter beside the wheelhouse for tailing prawns. Her side is sheathed to protect it from dredges and there is a power block aft for handling her trawl.

British Motor Fishing Vessels

SCOTS PAIR TRAWLER

Pair trawling for herring took off during the 1960s and drifting and ring netting eventually became redundant. To begin with drifters and ringers were adapted for pair trawling, but soon a very powerful series of wooden trawler/seiners was built for East Coast owners and began to make their mark on the herring, and later the mackerel grounds. As each ring netter had a neighbour, so the pair trawlers worked in groups. When a shoal was found, one boat would shoot her trawl and her partner would come and tow with her. When they completed the tow, the partner boat would make a tow with another of the team. Their skippers kept in touch by radio and no time was wasted. Among the well-known groups was the Peterhead 'Big Three' of the *Starcrest* PD 114, *Faithful* PD 67 and *Accord* PD 90. Another was the equally famous 'Big Five' of the *Fairweather V* PD 157, illustrated above, *Juneve III* PD 215, *Shemara* PD 78, *Sparkling Star* and *Ugievale II* PD 105.

In 1969 Macduff Boatbuilding completed the stylish yellow wooden trawler *Starella* PD 112 for Skipper Thomas Milne and Bruce's Stores of Aberdeen. The *Starella* was 78 feet in length with a 23-foot beam and an 11-foot draught. Her main engine was a Caterpillar 425 hp and there was a Lister auxiliary, which provided hydraulic power for her trawl and seine winch and power block aft. Her deckhouse, wheelhouse and spars were of steel. Wheelhouse equipment included an automatic pilot, watch alarm, radar, echo sounder, radio telephone, VHF and intercom. The *Starella* was to pair for herring with the *Achilles,* which was also distinctively painted yellow.

Forbes of Sandhaven delivered the equally well-equipped wooden trawler *Aquarius* FR 55 for Skipper Jim Slater of Rosehearty. Powered by a Stork 500 hp engine, the transom-sterned *Aquarius* was 77 feet 9 inches long with a 22-foot 8-inch beam and an 11-foot draught. Unusually, she had two four-berth cabins either side of a central mess room aft. Her fishroom could carry 500 crans of herring.

In 1972 Forbes delivered the powerful 80-foot cruiser-sterned *Kallista* FR 107. Driven by an 850 hp Caterpillar engine, she was capable of a speed of 12.5 knots. A Carron power block helped haul her trawl.

While powerful wooden trawlers continued to be built in the early 1970s, many pair trawler skippers opted for steel, among them Skipper John Alec Buchan and partners of Peterhead, whose *Fairweather V*, built at Faversham, was one of four Croan-class trawlers. This powerful vessel was fitted with a net drum aft. Instead of trawl gallows, she towed from extensions on either side of her deckhouse. She was 84 feet 11 inches long with a 22-foot beam and a 10-foot draught. Her main engine was a Mirrlees Blackstone of 637 hp.

Many pair trawl skippers wanted standard steel boats, of the same power, to make towing more straightforward. The class designed by Tynedraft of Newcastle, after detailed discussions with Skipper Jim Pirie and other Peterhead fishermen, included the *Unity* PD 209, *Juneve III* PD 215, *Shemara* PD 78 and *Morning Dawn* PD 195. These craft were 86 feet long with 22.2-foot beams and depths of 9.7 feet. Several were built by the John Hepworth yard at Paull on Humberside. Others came from Cubow at Woolwich on the Thames. They were built in the mid-1970s.

Another successful team were the identical *Bracoden* BF 37 for Skipper G. Alexander of Gardenstown, the *Steadfast Hope* FR 43 for Skipper J. Watt of Fraserburgh, and the *Replenish* for Skipper J. West of Gardenstown, all of which were built by the Hugh Maclean yard at Renfrew in 1970. These round-bilged vessels were 82.5 feet long with 22.5-foot beams. Others were built by Maaskant in Holland. Many steel boats were later lengthened and fitted with RSW tanks and shelter decks.

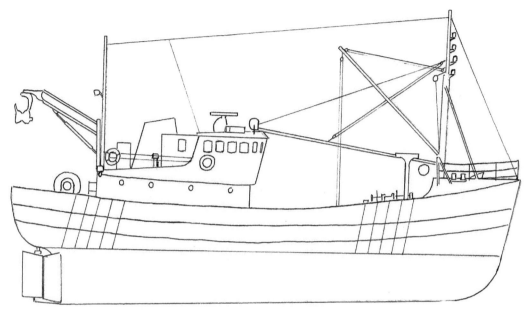

Scots pair trawler.

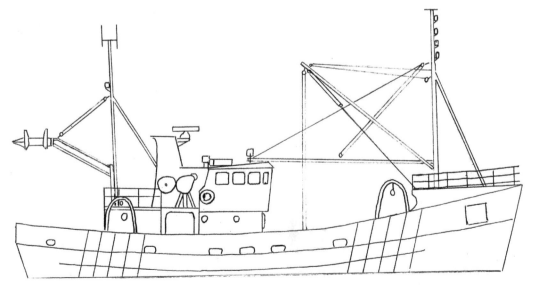

Tynedraft pair trawler.

British Motor Fishing Vessels

PURSE SEINERS

The purse seine is a large wall of net with floats at the top and rings at the bottom that is shot around a shoal of fish to encircle it. A wire or rope running through the rings at the foot of the net is hauled in to close, or purse, the bottom of the net, which is hauled in until the fish are alongside the boat. The catch is dipped out with a brailer or pumped out with a fish pump. Purse seining originated in the eastern USA, where mackerel schooners worked them from speedy rowing boats and dories. It was copied on the US and Canadian West Coast, in South America, in Portugal and the Basque Country (where steam sardine pursers worked the bolinche), and is still used by Brittany's thriving sardine fleet, and in Norway for herring fishing. Until power hauling arrived, pursers needed large crews to handle the massive nets.

The arrival of the power block and light strong artificial fibre nets transformed pursing. The power block was a rubber covered powered roller hung from a spar. Modern triplex haulers have three rollers. The power block enables very large nets to be worked in the open sea. The invention of the sonar, derived from Second World War ASDICs for detecting submarines, enabled skippers to accurately locate the fish. In the mid-1960s, the Norwegian purse seiner fleet demonstrated its ability to massively outfish the Scots drifter and ring net fleet.

Go-ahead Scots skippers began to build purse seiners. The earliest Scots-built wooden pursers were immensely strong, like beefed-up drifters. Many owners ordered steel vessels from Norwegian yards, which had years of experience building purse seiners. Others came from Maaskant in Holland and from the Clyde. The pursers rapidly developed. The earliest pursers boxed their fish but it was soon realised that chilled seawater tanks (CSW) or refrigerated seawater tanks (RSW) kept large catches at a much higher quality. The early pursers had skiffs to help them handle their nets or partner boats, which were often former drifters – for example, the large steel *Claben* worked with the wooden drifter *Anchor of Hope* – but they were soon fitted with side thrusters to keep them out of their nets. Year by year there were new developments, bigger steel vessels that could carry two seines, the fitting of RSW tanks and shelter decks, more advanced sonars and electronics, and ships being cut in half and lengthened.

Where the drifters' and ringers' landings were measured in crans, the pursers landed hundreds of tonnes. Old fishermen used to declare, 'You can't catch them all!', but the pursers soon demonstrated that they *could* catch them all. Quotas were imposed on landings.

By the end of the twentieth century, the purse seine fleet was reduced to a very few huge pelagic tank ships based in Shetland, Fraserburgh, Peterhead and Killybegs. Huge midwater trawls have replaced purse seines. This small group of expensive and high-tech industrial ships now catch most of the mackerel and herring landed in the British Isles. They spend much of the year laid up, waiting for seasons to begin.

Shetland purser *Zephyr* LK 319. The wooden *Zephyr* was built by Forbes in 1976 for Skipper J. Irvine; she was later fitted with CSW tanks and a shelter deck. In 1980 she was replaced by a larger steel *Zephyr* LK 394 from Norway, which was subsequently lengthened. (Photograph Glyn Richards)

The Kilkeel purser *Spes Magna* N 247. The steel *Spes Magna* was built in Holland in 1976 for Skipper R. Mc Cullough. She was later lengthened, fitted with RSW tanks and given a shelter deck. (Photograph Glyn Richards)

British Motor Fishing Vessels

SCOTS TWIN RIG TRAWLER

In the 1980s the Danes developed twin rig trawling in the North Sea. Two trawls are towed from three wires, and the trawls are spread laterally by doors – one on either side of the rig. The middle wire goes to a roller weight called the clump. This very efficient fishing method was quickly copied by the Scots fishing alongside the Danish fishermen. This resulted in a large fleet of Scots twin riggers, which mainly fished for prawns in daylight.

Fine wooden twin riggers were built until the end of the twentieth century, but all recent boats are steel. A modern twin rigger has her deck completely enclosed by a shelter deck (which is raised aft) for her net drums, which wind her nets aboard. Her trawl winch is forward and her warps run aft over the top of the shelter deck. Her nets are hauled aft, but the cod end is taken forward and lifted aboard over a hatch on the starboard side of the shelter deck, as can be seen in the photograph taken aboard the *Crystal Sea* SS 118.

The 70-foot twin rigger *Excel* BF 110 was built of wood by Macduff for Skipper John Watt of Gardenstown in 1998. She was equipped with a 705 hp Mitsubishi engine. In 2003 she was replaced by a 55-foot steel boat, which was also named the *Excel* BF 110. In 2010 a third *Excel* BF 110 was built by Macduff Shipyard for the same owner, this time being 62 feet long with a 508 hp Mitsubishi engine.

In 1998 Macduff Shipyard built the wooden twin rigger *Aspire* FR 793 for John Duthie of Fraserburgh. In 2002 she was sold to Andrew Buchan of Peterhead and renamed *Favonius* PD 17. In 2009 she was bought by Portknockie owners and renamed *Moray Endeavour* BCK 17. The *Moray Endeavour* is 69.5 feet in length with a 22.5-foot beam and is powered by a Caterpillar 480 hp engine.

Scots twin rig trawler.

Hauling the trawl in the *Crystal Sea* SS 118. The second *Crystal Sea*, she was also constructed by Macduff in 1989 as the 68.7-foot *Endeavour* BF 326 for David Lovie of Whitehills. In 1993 she was sold and renamed *Enterprise* BF 326. In 1999 she became the *Good Design* BF 151, before being sold to Newlyn in 2007, again as the *Crystal Sea*. She has since been replaced by a steel vessel also named *Crystal Sea*. (Photograph Skipper David Stevens)

The twin rigger *Bracoden* BF 37 was built of steel by Macduff shipyard in 1991 as the *Solstice* for Whitehills owners. In 1997 she was sold to Shetland and was acquired the following year by Sandy Alexander of Whitehills, being renamed *Bracoden*. She is 68 feet long with a 22.6-foot beam; her engine is a Caterpillar 425 hp. In her stern are hatches for her twin trawls, with the track for the clump between them.

British Motor Fishing Vessels

ORKNEY CREELER

Orkney has a long tradition of clinker boatbuilding. The Orkney yole was the local traditional clinker boat. In the 1960s, local boatbuilders developed a class of lobster boat that had the wheelhouse forward. The engine was beneath the wheelhouse, leaving the whole of the rest of the boat free for creels. These boats set the standard for creelers all around the Scottish Highlands and Islands.

In 1964 Andersons of Stromness completed the *Mayflower* K 852, which had a raised foredeck and forward wheelhouse. With a length of 34 feet 2 inches, a 12-foot 4-inch beam and a depth of 5 feet, she drew 3 feet 6 inches aft and was propelled by a Lister 36 hp air-cooled engine. The *Mayflower*'s hauler was belt driven from the engine and she was intended to work 150 creels. In 1965 Andersons built their second boat of this type, the 34-foot *Zinnia* K 887. Built of larch on oak frames, she was 34 feet long with a 12-foot beam. A 48 hp air-cooled Lister diesel gave her a speed of 9 knots. With a large forward wheelhouse, the *Zinnia* had wide side decks with a coaming around her aft working deck. Her forward cabin had accomodation for three crew.

The following year, they launched the 36-foot *Clamhan Dubh* K 898 for G. Richardson. Like several of the firm's boats, she was designed by Ewing McGruer of Edinburgh. With a 12-foot beam and a 3-foot 6-inch draught, the *Clamhan Dubh* had accomodation for four beneath her massive forward wheelhouse. Her hull design enabled her to achieve a good speed of 12 knots from her 100 hp Lister diesel. Intended for lobster and line fishing, she had a Ferrograph Offshore echo sounder to help find her grounds. Of a similar type was the 36-foot *Girl Wilma* K 923, which was built in 1967 for Mr Mainland of Ronsay. She was built of larch planks on oak and Canadian rock elm frames and was powered by a 75 hp Volvo Penta diesel, which gave a speed of 9½ knots. Her winch was locally made in Stromness and she was equipped with a Simrad echo sounder and Curlew radio telephone.

James Anderson paid tribute to the boat designer by naming the *Ewing McGruer* after him. She differed from previous boats in being carvel-built of larch and oak planks on oak frames. It was stated that she was designed to stand up to the rigours of a force 10 storm. The *Ewing McGruer* was 38 feet long with a moderate 12-foot beam and 3½-foot draught, and her Volvo Penta diesel gave her a speed of 14 knots. She had a distinctive hull shape, with a rounded forefoot and fine run aft to her square stern, which no doubt contributed to her remarkable speed. She was very well equipped with an American Hydro Slave winch, Decca Radar and radio telephone.

The 36-foot *Killholm* K 34, built by James Duncan of Burray in 1970, was one of twenty new boats built with the assistance of the Highlands & Islands Board. Other similar 36-foot clinker-built creelers were the *June Rose* CY 50 and *St Vincent* CY 7 for Barra from James Duncan, the *Boreray Isle* UL 105, and the *Viking* YH 461, which was built by J. W. Mackay of Stromness for Great Yarmouth. These were both old established Orkney boatyards. These Orkney boatbuilders also fitted out fibreglass hulls on similar lines to their wooden creelers, which became popular all around Scotland.

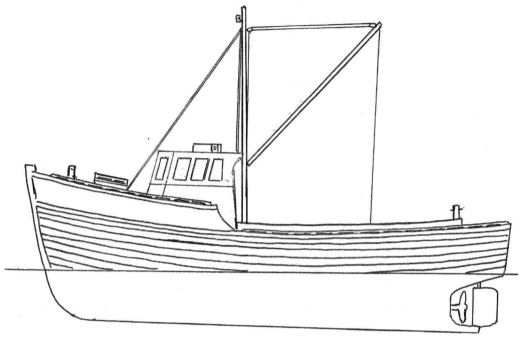

Orkney creeler.

Her Polperro owners were so impressed with the Orkney design that they had their long-liner *Patsy Anne* FY 23 carvel-built there in 1969. Constructed by J. T. Anderson of Stromness, the *Patsy Anne* was 38 feet long with a 12-foot beam and a moderate 3-foot 6-inch draught. Powered by a 135 hp Volvo Penta, she was fitted with a Simrad echo sounder and Sailor radio telephone.

MORECAMBE BAY PRAWNER/LANCASHIRE NOBBY

The Lancashire nobby, or Morecambe Bay prawner, designed to beam trawl for shrimps along the narrow channels off the Lancashire coast, was one of the few British working boats directly influenced by yacht design. The strong demand for shrimps by customers in the newly growing holiday resorts and the industrial towns of Lancashire led to the construction of these sleek cutters, with their distinctive rig and hull form, which was fully developed by the late 1890s. The forefoot was cut away to a rockered keel, the stern was an elliptical counter and the freeboard was low to enable the beam trawl to be more easily hauled. The nobby was fully decked with a long narrow cockpit surrounded by a coaming. A 35-foot nobby had an iron keel of about 1½ tonnes and another 3 tonnes of inside ballast. She set a cutter rig on a pole mast. Her catch was boiled aboard. Well-known builders of nobbies were Crossfield of Arnside, as well as Gibson and Armour of Fleetwood. They were to be found from the Solway Firth down to Cardigan Bay.

With the arrival of motors, they continued to work until the second half of the twentieth century. Their rig was reduced and they were fitted with wheelhouses at the fore end of the cockpit and capstans on the foredeck for hauling their trawls (*above*). The counter stern was vulnerable and often damaged; this was sometimes dealt with by simply sawing it off to leave a flat transom. Many nobbies have since been authentically restored as yachts

The traditional nobby continued to be built until the 1930s. In 1932 Crossfield built the 34-foot-long *Ann*. She had a 10-foot beam and a 3-foot 6-inch draught, and was powered by a single-cylinder Widdop diesel, which also drove the capstan. Though a motor boat, she carried a full sailing rig, including a bowsprit, and had no wheelhouse.

The shrimper *Isabella* BA 183 arrived from James N. Miller of St Monans, for J. Woodman of Annan, in 1962. Powered by a Gardner 70 hp diesel, she was 36 feet long with a 12-foot 6-inch beam and had a fifer capstan. Her forward wheelhouse was set down in the cockpit with the foremast in front of it.

In 1964 J. Harrison of Amble constructed the transom-sterned *Girl Helen* LR 123 for Sam Baxter of Morecambe Trawlers. A complete departure from the traditional nobby, she was 30 feet long with an 11-foot beam and 3½-foot draught, and was driven by a 36 hp Lister air-cooled engine beneath the wheelhouse. She had a spacious wheelhouse and a tripod foremast, which was an innovation at the time. The *Girl Helen* towed a 19½-foot beam trawl and was equipped with a Ferrograph echo sounder.

More recent shrimpers largely followed this lay out. Since fibreglass hulls have become available, they have been adapted for shrimping. In 1972 two 30-foot Tyler Hardy GRP hulls were fitted out for shrimping and white fishing in Morecambe Bay by the local Kellet and Milner yard. The *Mark Anthony* LR 55 for Charles Overett of Morecambe had a 10-foot 3-inch beam and a depth of 3 feet 3 inches. Her engine was a Lister 41 hp and she hauled her trawl with a Smallwood hydraulic capstan. Her shrimp boiler was fuelled by Calor Gas. The *Mark Anthony* was of similar layout to the *Girl Helen* but had a conventional mast and derrick. The almost identical shrimper *Kingfisher* was fitted out for Edward and Frank Gerrard. Her engine was behind the wheelhouse instead of below it, giving her crew a quieter life.

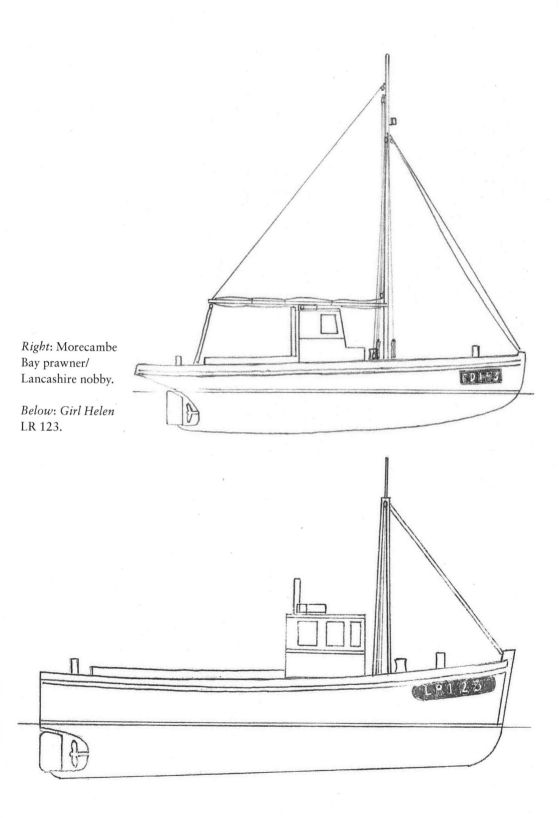

Right: Morecambe Bay prawner/ Lancashire nobby.

Below: *Girl Helen* LR 123.

British Motor Fishing Vessels

MANX SCALLOPER

The Isle of Man was famous for its herring fishery, which brought fleets from Scotland and Ireland to join local vessels. Manx kippers had a high reputation

The abundant queen scallops, caught locally, were once used only as long-line bait, but since the 1960s the island's fisheries for queenies and the larger king scallops have developed. While king scallops, which filter feed from a depression in the seabed, are caught with dredges, queenies can be trawled from June to October, as they leave the seabed to feed. The queenie fishery developed during the 1960s, when much of the catch was exported to the USA. In 1971 sixty vessels landed 7,500 tonnes. During the 1980s, new markets were found in France, Italy and Spain. In 2011 Manx queenies were given the Marine Stewardship Council's sustainability award.

The success of the Manx scallop fisheries depends on careful control by the island's government, which regulates the hours and seasons when fishing is permitted, the numbers of dredges worked, which grounds can be dredged, the number of boats with scallop licenses, and the allowable catch. In 2015–16, 4,500 tonnes of king scallops were landed, and in October 2016 the Manx government decided to limit the number of licenses to fish king scallops.

At present, a variety of vessels have been adapted for scalloping, and all dredgers are beam scallopers. When the fishery took off in the 1960s, Scottish boatyards constructed several boats with the wheelhouse forward, giving an unobstructed deck aft for working dredges (*above*). At that time, and before spreaders (beams) arrived, boats typically towed three dredges a side.

Among these elegant wooden cruiser-sterned boats were the *Manx Maid* CT 19, which was built by Thompson of Buckie for W. C. Watterson of Port Erin in 1958, the 45-foot *Rebena Belle* for J. T. & A. Gregeen of Port St Mary and the 50-foot *Fenella Ann* CT 27 for J. Cunningham of Port Erin (both of which were built by Alexander Noble of Girvan in 1960), the *Village Maid* CT 51 from Herd & Mackenzie for W. C. Watterson in 1961, and the *Heather Maid* CT 81, which was built by James Noble of Fraserburgh for F. & J. Watterson in 1965. Other classic wooden Scots boats fishing Manx scallops have included the *Maureen Patricia* CT 76, *Marida* DO 37, *Silver Viking* PL 19, *Coral Strand* PL 80, *Frey* CT 137 and *Constant Friend* PL 168.

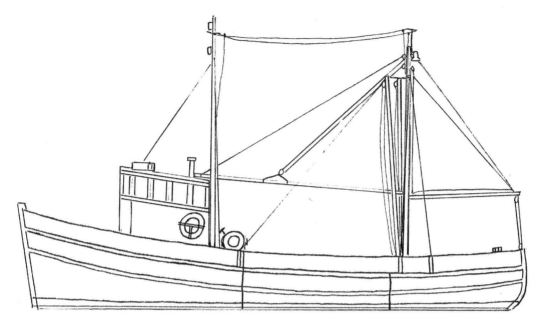

Manx scalloper.

Heather Maid CT 81 with her queenie trawl hauled up. (Photograph M. Craine)

BIBLIOGRAPHY

PERIODICALS

Chasse Marée
Commercial Fishing
Fishing News
The Mariner's Mirror
The Motor Boat
The Motor Boat and Yachting

REPORTS

Parliamentary Report on the Crab and Lobster Fisheries, 1877.
Parliamentary Report into the Application of Devon and Cornwall Sea Fisheries Committees for Grants from the Development Fund, 1913.
The Scallop and its Fishery in England and Wales MAFF Lowestoft, 1980.

BOOKS

Butcher, D., *The Driftermen* (Tops'l Books, 1979).
Butcher, D., *The Trawlermen* (Tops'l Books, 1980).
Drummond, P. and Henderson, S., *Fishing Boats of Campbeltown Shipyard* (The History Press, 2009).
Drummond, P. and Henderson, S., *Built by Nobles of Girvan* (The History Press, 2010).
Drummond, P. and Henderson, S., *Sputniks and Spinningdales* (The History Press, 2011).
Drummond, P. and Henderson, S., *The Purse Seiners* (Krohn Johansen Forlag AS)
Finch, R., *Sailing Craft of the British Isles* (Collins, 1976).
Finch, R. and Benham, H., *Sailing Craft of East Anglia* (Terence Dalton, 1987).
Hameeteman, C., *Britse, Schotse and Ierse Visserij in Beeld 1990–1999* (Lanasta, 2007).
Hawkins, L. W., The Prunier Herring Trophy (Port of Lowestoft Research Society, 1982).
Hawkins, L. W., *The Ocean Fleet of Yarmouth* (1983).
Hawkins, L. W., *Early Motor Fishing Boats*, (1984).
Lee, K., Stibbons, P. and Warren, M., *Crabs and Shannocks: Longshore Fishermen of North Norfolk* (Poppyland Publishing, 1983).

Macdonald, B., *Boats and Builders: The History of Boatbuilding around Fraserburgh* (Y. MacDonald, 1993).

March, E. J., *Sailing Drifters* (Percival Marshall, 1952)

March, E. J., *Sailing Trawlers* (Percival Marshall, 1953).

March, E. J., *Inshore Craft of Great Britain*: Vol 2. (David & Charles, 1970).

Martin, A., *The Ring Net Fishermen* (J. Donald Publishers, 1981).

Mitchell, P., *A Boatbuilder's Story* (John Gary Mitchell, 2014).

Nicholson, J. R., *Shetland Fishing Vessels* (*The Shetland Times*, 1981).

Peak, S., *Fishermen of Hastings* (1985).

Pengelly A. J., *Oh for a Fisherman's Life* (Glasney Press, 1979).

Potter, T. J., *Fishing with the Ropes* (2012).

Staley, A., *Last One Down the Slip* (Whitstable Museum and Gallery, 2005).

Sutherland, I., *From Herring to Seine Net Fishing on the East Coast of Scotland* (Camps Bookshop, 1985).

Tarvit, J., *Steam Drifters: A Brief History* (St Ayles Press, 2004).

Thomson, D., *Pair Trawling and Pair Seining* (Fishing News Books, 1978).

Thompson, D., *Seine Fishing* (Fishing News Books, 1981).

Weatherhead, F., *North Norfolk Fishermen* (The History Press, 2011).

White, M. R., *Fishing with Diversity* (2000).

White, M. R., *Herrings, Drifters and the Prunier Trophy* (2006).

Wilson, G., *Fishing Boats of Whitby and District* (Hutton Press, 1998).